# Manufacturing and Design

# Manufacturing and Design
## Understanding the Principles of How Things Are Made

### First edition

**by**

**Erik Tempelman**

**Hugh Shercliff**

**Bruno Ninaber van Eyben**

AMSTERDAM • BOSTON • HEIDELBERG • LONDON
NEW YORK • OXFORD • PARIS • SAN DIEGO
SAN FRANCISCO • SYDNEY • TOKYO

Butterworth-Heinemann is an imprint of Elsevier

ELSEVIER

Butterworth-Heinemann is an imprint of Elsevier
30 Corporate Drive, Suite 400, Burlington, MA 01803, USA
The Boulevard, Langford Lane, Kidlington, Oxford OX5 1 GB, UK

First edition **2014**

**Notice**

No responsibility is assumed by the publisher for any injury and/or damage to persons or property as a matter of products liability, negligence or otherwise, or from any use or operation of any methods, products, instructions or ideas contained in the material herein. Because of rapid advances in the medical sciences, in particular, independent verification of diagnoses and drug dosages should be made.

**British Library Cataloguing in Publication Data**
A catalogue record for this book is available from the British Library

**Library of Congress Cataloging-in-Publication Data**
A catalog record for this book is available from the Library of Congress

For information on all **Butterworth-Heinemann** publications
visit our web site at store.elsevier.com

14  15  16  17  18      10  9  8  7  6  5  4  3  2  1

ISBN: 978-0-08-099922-7

Working together
to grow libraries in
developing countries

www.elsevier.com • www.bookaid.org

# Contents

# Acknowledgements

This book has evolved from the course of the same title in the Industrial Design Engineering program at Delft University of Technology. We are indebted to the following colleagues and guest lecturers who have contributed to the course reader and the compilation of the book.

*Major contributors*
Em. Prof. Laurens Katgerman, Delft University of Technology (Chapter 3)
Jos Sinke, Delft University of Technology (Chapter 4)
Wim Sillekens, European Space Research and Technology Centre (Chapters 5-6 and Section 13.2)
Job Kneppers, Kneppers Ontwerp en Realisatie BV (Chapters 7 and 12)
Em. Prof. Jan Spoormaker, Delft University of Technology (Chapter 8)
Cor Janssen, DUROtherm Kunststoffverarbeitung GmbH (Chapter 9)
Kasper Snijder, Infinious BV (Chapter 10)
Oscar Brocades Zaalberg, BPO BV (Chapter 11)
Eric Biermann, Van Berlo BV (Chapter 12 and Section 2.3)
Prof. Peter Rem, Delft University of Technology (Chapter 14)

*Other contributors*
Chris Nijkamp, Fabrique Public Design BV (Section 2.2)
René Hop, Koninklijke Nedschroef Holding BV (Section 6.6)
Huub Ehlhardt, Philips Innovation Services (Section 8.10)
Willy Froon, W. Froon Consult BV (Section 8.11)
Remco Boer, Protyp BV (Section 8.12)
Zjenja Doubrovski, Prof. Jo Geraedts, Bram de Smit, Jouke Verlinden, Delft University of Technology (Chapter 11)
Erik Thomassen, Delft University of Technology (Sections 13.3 and 13.9)
John Powell, Laser Expertise Ltd (Section 13.4)
Ruurd Pieter de Boer, Enitor BV (Section 13.6)
Hanneke Agterhuis, Van Gansewinkel Group (Section 14.6)
Floris-Jan Wiegerinck, Marin Licina, Salar Vakili (Figures)
Dorothee Pape, Studio Ninaber (Cover design)

Our thanks to Mike Ashby and Marc Fry of Granta Design, Cambridge, UK, for permission to use data extracts from the Cambridge Engineering Selector software. We also thank the following colleagues who have taught the material at Delft, contributing through the exercises they have set, and their general advice and suggestions: Martien Bakker, Bert Deen, Henk Crone, Sebastiaan van den Elshout, Stefan van de Geer, Anton Heidweiller, Jan Willem de Kruif, Gerard Nijenhuis, Roald Piera, Joost Prins, Zoltan Rusak, Bram de Smit, Servaes Spiekerman and Erik Thomassen. Finally, thanks go to the hundreds of students at Delft University of Technology, whose feedback on the course has guided the refinement of the book's contents.

Erik Tempelman, Hugh Shercliff, Bruno Ninaber van Eyben
Delft/Cambridge, February 2014

# Preface

## TEACHING MANUFACTURING AND DESIGN

In teaching manufacturing, the classic starting point is to group all processes into categories, starting with their primary functions (shaping, joining, surface treatment). Shaping processes tend to be subdivided by material class (metal, polymer, etc.) or the material state (solid, liquid, powder): hence, metal casting, metal forming, polymer molding, and so on. All processes within one category then have certain aspects in common, making understanding easier. However, for teaching and learning, this approach has its limitations if applied too rigorously. On the one hand, processes within one category can still be worlds apart when it comes to putting them in a *design context*, technically, economically, or both. On the other hand, there are useful parallels between processes that fall superficially into different categories, meaning that educational opportunities can be missed—for example, at the level of the mechanism of metal removal, machining has things in common with metal forming. For similar reasons, categorizations are poor at dealing with new processes: for instance, thixomolding lies somewhere between deformation and casting, whereas additive manufacturing, also known as "3D printing," is something else altogether. In summary, categorizing processes is only a starting point for teaching and learning about manufacturing and design.

This book provides an alternative, based on a division between the hierarchical levels of the *principle*, the *method*, and the *equipment* used for manufacturing, and based on a distinction between the attributes of processes with respect to *function*, *quality*, and *cost*. This structuring of thought processes around manufacturing has proven to be very useful for a design setting, especially because it invites a gradual choice for manufacturing processes during the product design process, moving steadily down the hierarchy while going from conceptual design to product detailing—all the time striking trade-offs between attributes of function, quality, and cost in all their forms.

A second novel aspect of the book is its selective nature, as it covers only those processes that the suggested readers are most likely to encounter during their careers. This makes it very different from the conventional "encyclopedic" textbooks—and, consequently, much more manageable. The book also gives guidelines for how to build up knowledge on the processes not covered here, encouraging a philosophy of lifelong learning.

Furthermore, we have opted to address the "how" of the selected processes just as much as the "what" by presenting the underlying physics and material science. This is something that works especially well at the level of the manufacturing principle, although again we have been selective by focusing on the main concepts and terminology. Mixed with the right respect for practical considerations, this science-led approach allows the reader to distinguish what is (im)possible in principle from what is (im)possible in practice, and this is a vital skill for professional designers in their interaction with the manufacturing industry. It also invites readers to explore generally applicable concepts, to discover ways to make simple models of complex phenomena, and to forge a strong connection between manufacturing and materials science.

The book has a final innovation: exercises do not appear the end of chapters but are embedded in the main text itself, at strategic locations, and they are deliberately open-ended. Furthermore, full solutions are not provided, although hints and examples are (in the online appendix). This setup might seem strange at first sight, but it has already proven itself to be effective in encouraging students to become active learners. Overall, we expect that professionals, teachers, and students alike will appreciate the novel approach to teaching manufacturing processes for product design that is taken in this book.

## READERSHIP

This book is written primarily for bachelor's-level students of industrial design, design engineering, and general mechanical engineering (including manufacturing, materials, aerospace, and automotive), as well as their teachers. It will also prove to be worthwhile for students of other disciplines relating to design and engineering, as well as for design art school students who are not put off by a few equations. Basic high school physics and mathematics are required, as is some acquaintance with materials science. *Manufacturing and Design* can also be recommended for people working as professionals in the fields of materials, manufacture, or design, for whom it may serve as a fresh-up course or even as inspiration for solving production-related design problems.

## THE CHANGING WORLD OF MANUFACTURE

Until the 1980s or so, dealing with manufacturing during product design was comparatively easy. Most designers worked either for original equipment manufacturers (OEMs) that had in-house manufacturing facilities or for design agencies that had, through their OEM clients, strong ties to those facilities. So all necessary knowledge and experience could be found either in-house or right around the proverbial corner. Moreover, the manufacturing specialists were "on the same team" as the designers. Apart from those organizational and communicational benefits, the world of manufacturing processes was not changing too rapidly, and product life cycles were comparatively long, leading to design processes we would now consider to be relaxed.

Today's designers have no such luxuries. Many OEMs have outsourced their manufacturing facilities, often to faraway countries, and knowledge and experience are hard to come by. And when designers finally meet the specialists, these individuals are no longer on the same team and now have all the reason to use the difference in knowledge to their advantage. Development times are short, leaving little opportunity to reflect. Even more crucially, the world of manufacturing is changing faster than ever, making it ever more likely that real chances are missed to deliver the same functionality with better quality, or against lower cost, or to find another road towards market success. Any one of these factors alone forces designers to reconsider what they think they know about manufacturing, let alone the force of all of them together!

This book presents a way out. By tackling the vast and dynamic field of manufacturing processes from a fresh, different angle than existing approaches, as well as by being very selective, it allows you to make the most out of the increasingly limited time that students, teachers, and professionals have available. It gives designers the knowledge needed to not only create optimal designs, but also to understand where the supplier is coming from, so that together, they can search through the manufacturing options and find the best compromise.

Ultimately, this book aims to raise our level of thinking and take manufacturing and design into the 21st century. As long as one does not forget about practicalities, understanding the physics behind the processes is key to understanding what is possible and what is not, but for too long this topic has been ignored or made incomprehensible in university textbooks. *Manufacturing and Design* aims to change this situation.

# Introduction

## CHAPTER OUTLINE

## 1.1 MANUFACTURING: THE ROLE OF THE DESIGNER

All products you see around you, from the simple disposable coffee cup in your hand to the hugely complex aircraft overhead, have been *designed*. They have also been *manufactured*. Without exception, successful mass-produced products have been designed to be not just functional and pleasing to use but also affordable to make, striking the right balance between the conflicting product requirements of functionality, quality, and cost. In the design process, whose role is it to ensure that products *can* in fact be manufactured well? Before answering that question, let's look more closely at what it involves.

In design for manufacture, the first major decision typically concerns the overall structure of the new product—its layout, or architecture if you wish. The iterative process of choosing the material starts soon afterwards. Next comes the 'make-or-buy decision': deciding which components to buy in from stock and, conversely, which parts to design specifically for the new product. (Increasingly, products share components among different versions as well as generations, and this complicates both the choice of product layout as well as the decision to make or buy.) The product-specific parts are then designed in detail, which involves choosing the right manufacturing processes and is intimately linked to material selection. This is also the time to consider assembly: how will all parts and components be joined together into a working product? More often than not, it will become apparent along the way that certain design choices do not combine well with decisions made earlier. The solution is to trace back one's steps and explore alternatives before converging onto a satisfying solution. That, in a nutshell, is what design for manufacture entails (Figure 1.1). It is a difficult, complex, and time-consuming process that nevertheless lies right at the heart of design. It can also be very rewarding, especially when choices for manufacturing processes not only realize but actually *improve* the initial design ideas.

So whose role is it to take on this challenge? That, dear reader, is *your* role. Regardless of whether you are an industrial designer, mechanical engineer, aerospace engineer, or similar creator, your job is to do design for manufacture, just as you should address functionality, aesthetics, and any other requirement that your product must meet. Suppliers will gladly help you to ensure that the parts you design are feasible to make, or that parts and components are easy to assemble. But selecting a supplier means

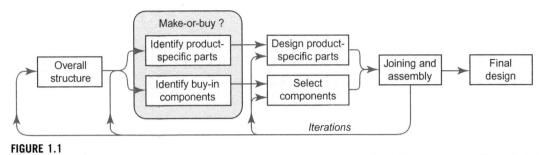

**FIGURE 1.1**

Key steps in design for manufacture. (For color version of this figure, the reader is referred to the online version of this chapter.)

selecting a process first, at least at the top level of the manufacturing principle—for example, choosing between injection molding or thermoforming for a plastic part, or between casting or forging for metals. Although suppliers know their own process very well, they often have a limited view of alternatives and a vested interest in their own. They cannot therefore be relied on to guide such process choices—that is your job. Also, they will happily help your competition as well, so if you listen too much to suppliers, your design will not stand out from the crowd. Finally, the manufacturing industry tends to steer you in the direction of what has been done before, not of what could one day be, resulting in conservative designs that may be easy to make but that are not innovative. How can you avoid this trap? The answer: by building up a solid background of what manufacturing processes can and cannot do. That is what this book will help you to do.

## 1.2 PRINCIPLES, METHODS, AND EQUIPMENT

Industrial manufacturing practice comprises a vast number of processes, each of which can be encountered in numerous variations. Every process—often every piece of manufacturing equipment that a supplier has available—has its own possibilities and limitations. Understanding requires in-depth knowledge, but how can you acquire this knowledge when the number of process variants is so large and your time for learning is always limited?

The first step of the solution provided by this book is careful *selection*. After consulting a number of design professionals from various fields and product categories, a list was drawn up of processes that—with few exceptions—every designer should aim to know and understand. Next, these selected processes were ordered into an effective *hierarchy*, bringing processes together that not only have strong similarities in terms of the underlying physics and material science but also represent distinct industry branches. This leads to the hierarchical level of the *manufacturing principle*, and that of the *manufacturing method*. As parts and products eventually need to be made on a real piece of machinery, a third and final level to identify here is that of the *manufacturing equipment*.

Figure 1.2 presents an example for the group of casting processes for metals. Sand casting, low-pressure (LP) die casting, and so on are all methods under the same principle, and underneath each of these methods are all kinds of specific equipment. What may be a limitation of certain machinery (e.g., size, accuracy) need not be a restriction for other equipment; conversely, there can be

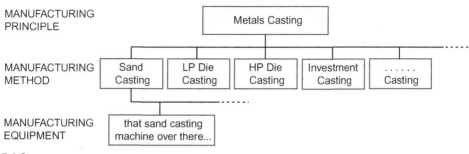

**FIGURE 1.2**

Levels of abstraction—principle, method, equipment—as applied to casting processes. (For color version of this figure, the reader is referred to the online version of this chapter.)

requirements that certain methods may find hard to meet or that are even inherent limitations of the principle itself.

To summarize, the resulting process hierarchy distinguishes among the following characteristics:

- What is possible in *principle* (as determined by physical law)
- What is possible in practice—that is, the *methods* under that principle (as determined by industrial state-of-the-art)
- What is possible with the *equipment* used for that method (as determined by what a certain supplier has available)

## 1.3 SCOPE AND CONTENT OF THE BOOK

This book deliberately does not aim to be comprehensive. Its focus, to start with, is on *shaping* processes for parts and products, with limited space for joining and assembly, and even less for finishing. So the emphasis is on product-specific parts rather than on standard bought-in components and semi-finished stock material. It also restricts itself to those mass manufacturing processes that you are most likely to encounter, with only limited attention to less common processes. Likewise, prototyping and one-off production are only addressed briefly, whereas some processing areas (such as fabrics and textiles, or the micro-manufacture of electronic devices) are not covered at all. So what you have in front of you is a *selective* book indeed.

Table 1.1 summarizes the contents of this book. Chapters 3 to 12 form the core of the book. Each of these chapters deals with one particular manufacturing principle, explaining how it works and how it can be described and understood in physical terms. Each chapter ends with hands-on information on the various methods commonly found under that principle, as well as pointers for further study. These chapters have been written as independently as possible from each other and can therefore be read in any order or number. The remaining five chapters complement the core, as follows. Chapters 1 and 2 are scene-setting chapters that introduce the book, show what several common products are made of, and identify which processes were used to make them. Readers are invited to perform such "product disassembly studies" for themselves and to reflect on why these particular materials and processes were used. Chapter 13 contains brief descriptions of 10 additional processes, including product finishing, as well as guidance for gathering information and building insight into processes not covered in this book

**Table 1.1** Manufacturing Principles and Methods Covered in This Book

| Ch. | Principle | Methods |
| --- | --- | --- |
| 1 | Introduction | |
| 2 | Product disassembly studies | |
| 3 | Shape casting of metals | Sand casting, gravity die casting, low- and high-pressure die casting, investment casting |
| 4 | Sheet metal forming | Bending, roll bending, roll forming, panel beating, deep drawing, matched die forming, sequential die forming, rubber forming, hydroforming |
| 5 | Extrusion of metals | Hot direct extrusion of aluminum, hot direct extrusion of other metals |
| 6 | Forging of metals | Open-die forging, closed-die forging, pressure forging, net shape forging, cold heading, coining |
| 7 | Machining | Turning, drilling, milling, integral machining |
| 8 | Injection molding | Compact molding, insert/outsert molding, gas injection molding, molding with in-mold decoration, 2K molding |
| 9 | Thermoforming | Free forming, standard forming, pre-stretch forming, plug-assisted forming, twin-sheet forming |
| 10 | Resin transfer molding (RTM) | Pressure-controlled RTM, volume-controlled RTM, resin infusion, light RTM |
| 11 | Additive manufacturing | All seven sub-principles as classified by the American Society for Testing and Materials (ASTM) in 2012 |
| 12 | Joining and assembly | Welding, soldering, adhesive bonding, mechanical fastening, joining using form closures, line versus cell assembly |
| 13 | None of the above | Ten additional processes |
| 14 | Recycling | |
| 15 | Manufacturing process choice | |

at all. Chapter 14 covers the basics of recycling, which is, in many ways, the reverse of manufacturing processes (products go in, materials come out). Finally, Chapter 15 draws many threads of the text together by summarizing the considerations for making well-balanced choices regarding manufacture during the product design process.

In the core chapters, most of the attention goes out to the principle. Once more, the book is selective, digging no deeper than strictly necessary. Each principle is explained by using a small number of key concepts and introducing only essential terminology. The aim is not tell you all there is to tell, but instead to focus on those things you need to know to make a difference. However, even with a limited discussion of the theory underlying the various principles and by showing how it can be applied to the methods under them, you will build up the right knowledge and insights—making the most of your limited time.

In some books, this physics-based approach can tend to distance readers from the actual nuts and bolts of the subject matter. Chapter 2 aims to prevent this from happening by showing instructive examples of the real world of manufacture, and inviting readers to literally take products apart themselves. Also, each core chapter starts with an exercise that encourages you to find and study parts made under each principle. With this initial exploration of products, the focus on underlying physics immediately has a context and shows its key benefits. Instead of keeping the "black box" of manufacturing closed and merely describing the "what" of a given process (as most manufacturing textbooks do), this approach allows you to look inside the box and discover the "how" and "why" as well—which is generally more satisfying for the inquisitively minded reader. This approach is also universal, capable of explaining important differences or similarities between various methods with insight and understanding. Another benefit is that it makes the underlying materials science come more alive by forging a strong and useful mental connection between materials and processes. (Indeed, practicing this approach with attention to simple process modeling and the correct use of quantities and units will also help you to understand other fields, not just manufacture and materials.) Finally, because the state of the art may change but physical behavior remains constant, it is also an approach that is future-proof.

Readers who want to move beyond the text are provided with references for doing so. A word of warning though: manufacturing information can appear encyclopedic—nobody can know all of it. What is most important is to develop sound thought processes that transfer readily to any scenario linking processing to materials and design. We therefore focus on structured thinking. Our hierarchy of principles, methods, and equipment comprises the first half of our structure; the second half is formed by our "manufacturing triangle," to which we turn our attention now.

## 1.4 THE MANUFACTURING PROCESS TRIANGLE

Every process has certain characteristic capabilities, which may be captured as numerical data ranges or lists of qualitative parameters. For example, the typical size range that a process can handle, or the materials that it can operate on, are clearly important attributes of the process. Matching the requirements of a design to suitable processes is at the heart of design for manufacture. So a database of process attributes would be a starting point, but is only viable if the data are aggregated across all materials and all design scenarios. Unfortunately, this has significant limitations, as the ranges of these parameters can end up very wide, weakening comparison and choice between them. This will become especially so on the level of principles, but already on the level of methods it becomes unwieldy, when the equipment for individual methods varies considerably as well. Moreover, these quantities are not independent of one another—for example, larger parts or stronger materials need more powerful equipment, operating more slowly. Achieving a successful design that can be made economically, to a sufficient quality standard, is therefore an optimization problem, involving many *trade-offs*. So it is crucial to recognize that the process attributes are not fixed quantities but combinations of variables, with important couplings and trade-offs between them.

Given the great diversity of materials, designs, and processes, this effort can quickly appear overwhelming. What is needed is a form of "knowledge management"—a structured approach to order our thinking in an unfamiliar design or processing scenario. To do this, this book defines and organizes attributes under three headings: *function, quality,* and *cost*; these categories form the corners of a "manufacturing process triangle." The triangle informs the competition between processes using different principles, such as: shall we cast this alloy or forge that one? Differences between manufacturing methods under one principle can then be explained via the different settings for certain key process parameters. Trade-offs can exist between the three corners of this triangle as well as inside them, and they can emerge on any of the three abstraction levels presented previously. So let's expand this manufacturing triangle, and define a structured thought process for the rest of the book.

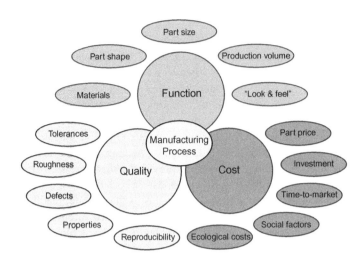

**FIGURE 1.3**

The "manufacturing triangle," showing the dominant process attributes relating to function, quality, and cost. (For color version of this figure, the reader is referred to the online version of this chapter.)

Figure 1.3 shows the full triangle, with the five key attributes associated with each corner. Convenient shorthand for the three corners could be as follows:

- *function* = can it be done?
- *quality* = can it be done well enough?
- *cost* = can it be done cheaply enough?

This triangle and the 15 inter-related attributes provide the structure for addressing all of the processes in the book. First we present the attributes themselves.

*Functional attributes* are the *materials*, *shape*, and *size* of the part to be produced (or the product, if it is fully finished). Then there is the *production volume* (i.e., the requirement to produce not one but a certain number of parts). This attribute originates mainly from process economics, but accumulated experience and data allow it to be considered as a separate functional attribute, without going into the actual costs. These four are well defined and can be objectively quantified. But often there are additional functional elements to consider that defy easy definition, relating to the "look and feel" of the part. This catch-all term captures the part's color, texture, and expression. Few processes can address all five functional aspects in one step—indeed, subsequent assembly and finishing (e.g., painting) will usually be required—but they all deserve the designer's attention.

The *quality attributes* start with *tolerance* and *roughness*—not to be confused with one another. Tolerances are the inevitable deviations to the part size that the process delivers, which can be large (wide tolerances) or small (narrow tolerances). Roughness on the other hand is several orders of magnitude smaller and refers to the part's surface, which can be smooth or rough. It is, of course, linked to look and feel. *Defects* include surface and bulk imperfections, usually small compared to the part, and of various types—cracks, pits, or surface markings. The next quality target is to achieve the desired material *properties*. Here it is vital to appreciate that processes not merely bring the material into the required shape, but that more often than not they simultaneously affect the final material properties.

This is particularly true of the key mechanical properties of strength and ductility—indeed, some dedicated processes, such as heat treatment, are designed purely to modify these properties. Finally there is *reproducibility*—that is, the issue of ensuring that parts are produced with consistent quality from one unit in the batch to the next. This captures the *variability* of the other quality attributes, as opposed to the average target value. As before, the first two attributes are well-defined and objectively measurable, whereas for the others there is considerable technical difficulty. The occurrence of defects, the material properties, and the variability of both depend in complex ways on the material, process, and design detail—but they all matter nevertheless.

Finally there are the five *cost attributes*. Foremost is the *part price*, which includes contributions from the materials and consumables used, the depreciation of all dedicated capital costs (such as molds and dies), transport and packaging, and time-related costs. The last of these is itself a complex hybrid of costs charged by the hour (labor, energy, overhead costs, plus a share of the equipment capital, which is usually amortized over a period of, say, five years). The relative importance of each cost contribution depends further on factors such as production rate, also known as "cycle time"—processes that are expensive to run look more attractive if they have a rapid throughput. Similarly, trade-offs between capital, labor, energy, and transport costs figure largely in decisions to transfer production around the globe. But one contribution is sufficiently influential that we list it as a separate attribute: the specific *investment* needed for making the part. This attribute can drive process choice all by itself, especially when the production volume is uncertain: no one likes to pay up front for an expensive mold for a product that may not sell as well as expected. For shaping processes at least, part price and investment can be well defined and quantified (e.g., in euros), but there are three additional cost attributes to consider that are more difficult to pin down. This includes *time-to-market*—that is, the time required to go from the initial design steps to the moment of market introduction—which can easily be months or, for complex products, years. It is a major driver in manufacturing process choice, with ever shorter product life cycles being demanded and all that this implies for managing the development costs (and risks) in product design and engineering. Finally, the hardest of all, there are the *social factors* and *ecological costs*—central to the rapidly evolving concept of sustainable development. This topic is of such importance that it merits its own consideration to set it properly in the context of this book.

Companies and governments increasingly approach the concept of national or global wealth as the sum of three components: *manufactured capital*, *human capital,* and *natural capital*, sometimes described as the "triple bottom line." The first of these components has dominated profit-driven design and manufacturing since the start of the Industrial Revolution, but a balance between all three has become the new imperative. It is driven by the consequences of rapid globalization and the fast-growing emerging economies, coupled to the evidence for man-made climate change. Some consideration of sustainability must now be central to all product design and manufacturing, but it is a fast-changing field. Educational approaches to the topic are developing all the time, from simplified but quantitative life cycle analysis of energy consumption or $CO_2$ emissions, to discursive investigation of the wider, highly diverse, factors that underpin assessment of the three capitals (the stakeholders in the product and its consequences, material and energy sourcing, ethical labor practices, regulation and legislation, public perception, and so on). A full coverage is well beyond the scope of this book. Here, sustainability will be addressed in a few specific ways. First, we note that the energy consumption in manufacturing a product is almost always a minor contribution compared to the embodied energy in producing the raw material or the energy consumed in the use phase of the product life cycle. Recycling processes are therefore very important, which is the main reason for including Chapter 14. But note also that much of the impact of sustainable design on processing is indirect. Design may seek to improve thermal, electrical, or mechanical efficiency, or to reduce weight for transport applications, or to improve the end-of-life material recovery. In all cases, this will lead to changes in the materials and shapes chosen for the design, with greater use of light alloys and composites, or seeking greater form freedom. So consideration of these attributes in processing may be a consequence of sustainability, but it does not in general require a change in our approach to manufacturing itself.

In summary therefore, the manufacturing triangle shown in Figure 1.3 captures the approach we will take to exploring each processing principle and evaluating trade-offs between attributes—and, indeed, between processes themselves. Of course, it is perfectly possible to make other classifications of attributes or to give more weight to some attributes over others. The triangle presented here is a tool, not a straitjacket. But the fact remains that some kind of trade-off must always be made when it comes to manufacturing process choice. The triangle provides a framework for exploring manufacturing and design—a simple checklist of things to think about. It is used consistently throughout the book, so you will encounter many examples of what the attributes mean for the various processes. The final chapter, on manufacturing process choice, again looks deeper into the triangle and explains which attributes matter most during the various phases of the product design process.

## 1.5 HOW TO USE THIS BOOK

*I hear and I forget—I see and I remember—I do and I understand.*

**Confucius, 500 BCE**

The book is at its most effective when it is used as the primary textbook for a dedicated academic course, embedded in a range of teaching and learning modes. For example, the way the material has been covered at TU Delft is as follows. Per chapter, students first receive a plenary lecture on key concepts and terminology. Videos are shown of the process in action, and sample parts are handed around to give students a first feel for the results. For homework, the student then spends *at least* three hours per chapter, working alone or in pairs. Step 3 is the discussion of this homework, in two-hour sessions, with one teacher per 15 students or so. In addition, factory tours are recommended at some time during the course: it makes the processes come alive, especially at the equipment level. There is nothing like seeing a real manufacturing environment, which may be fast, dynamic, hot, and noisy, to gain an appreciation of the practical limits to theory learned in the classroom. Finally, the knowledge acquired needs to be applied. This can be done through research projects in which students study a certain novel manufacturing process (applying Chapter 13), through a design project (applying Chapter 15), or both. Different institutions will have different constraints (not least financial), and, of course, the book could be simply read as self-study. But design and manufacturing are inherently tactile, hands-on activities, so learning is enhanced by handling real materials and parts and trying to (re-)design actual products.

The approach toward providing exercises in this book is quite different from what might be expected and deserves some explanation. First, the exercises are not placed at the end of each chapter but are strategically scattered across the main body of the text. This is done specifically to activate the reader, ensuring that the key concepts are explored before moving on. Often the questions can be answered purely by searching through the preceding text for answers, but they may also require further research. Students using this book are therefore recommended to have alongside both paper and pencil plus a laptop or other device to access the Internet. Finally, in another deviation from convention, the book deliberately does not include the answers to all exercises. This strategy aims to counter the pervasive "look-up" behavior that so often stands in the way of building up insight; apart from that, some exercises—notably those where students are asked to devise their own design guidelines—are simply too open-ended to allow full solutions. In the online resources, the reader will find plenty of hints (as well as occasional indications of order-or-magnitude answers) plus a set of fully worked-out, exam-style questions and answers.

Like every written source, this book will not work well for readers who are unwilling to work at the book themselves. Students will learn very little if they do not search for sample products, or do not think about the text and questions but immediately skip to the hints, or do not try to formulate their own design guidelines. There simply is no royal road to learning, and *Manufacturing and Design* is no exception. But for those readers who do get involved, this book should prove to be very useful indeed.

## 1.6 **Further Reading**

Ashby, M.F., Shercliff, H.R., Cebon, D., 2013. Materials: Engineering, Science, Processing and Design, third ed. Butterworth-Heinemann, Oxford, UK. (*A design-led text that starts from the design requirements for products, leading to material and process selection, and illustrating the links between processing and properties. Chapter 20 introduces a wider discussion of sustainability and design.*)

Miodownik, M., 2013. Stuff Matters. Penguin Books, London, UK. (*An excellent book on materials for the general public that acquaints you with many key aspects of materials science, covering traditional materials as well as "future stuff." Highly recommended as background material, for providing an alternative view, or simply for entertainment.*)

# Product Disassembly Studies

## CHAPTER OUTLINE

## 2.1 INTRODUCTION

This book aims to familiarize you with the main manufacturing processes, as well as their possible applications in all sorts of products. This require a hands-on approach, whereby you actively study the products you encounter in everyday life and ask yourself how they are made. Indeed, most chapters begin with an exercise designed to stimulate you to do precisely this: for instance, Chapter 3 on casting of metals asks you to identify several cast parts and products right from the start. It might be tempting to limit yourself to an image search on the Internet, or even to skip those exercises altogether, but that would be a mistake. It is important to have such products at hand, so you can not only see them from a distance and at just one angle, but look at them closely, from all angles, and literally get a feel for the component. This way, many of the theoretical and practical concepts covered in this book will come much more alive.

This chapter is intended to help you make these instructive explorations. Section 2.4 introduces you to product disassembly studies, which basically involve taking products apart, considering how their various components are made, and finding out how they have been assembled together. If you are technically inclined (which every student of design and engineering must be to some extent: otherwise, why choose such studies?), you have probably already taken apart quite a few products during your lifetime, and you can now learn how to take this hobby to the next level. One quick warning: *always* consider safety, and *never* put yourself, or others, at risk.

Before setting you off on disassembly of products, we show a few exemplars of everyday products and discuss how they were made. As you will see, nearly all of the manufacturing processes covered in Chapters 3 to 12 are encountered in these sample products. Of course, many more could have been included, but because this book aims to be selective, we limit ourselves to just two products: one outdoor application and one typical indoor product, both examples of industrial design.

**EXERCISE 2.1**

Find a product that is ready for recycling or disposal—for example, an old vacuum cleaner, coffee machine, or computer, or perhaps something bigger such as a lawnmower (preferably products that contain mainly metals and/or plastics). Save it for Section 2.4!

## 2.2 OUTDOOR DESIGN: BUS SHELTERS

Bus shelters come in all sorts and sizes. Over the past decades, their design and manufacture has developed to a degree of sophistication that may be surprising at first, but that becomes readily apparent to anyone who gives these objects a closer look—and considering that if you use them you inevitably spend some time in them, we suggest you do so. Here, we present a bus shelter known as "De Gouwe," designed for various municipalities and provinces in the Benelux region. It is currently manufactured by the company Epsilon Signs in Bree, Belgium, with placement and maintenance done by the company OFN in Buren, The Netherlands. Fabrique Public Design in Delft, The Netherlands, was responsible for its design. All information and illustrations in this section are courtesy of this design bureau, which has been active in bus shelter design since 1993.

Figure 2.1 shows an exploded view of one version of De Gouwe with its main components and how it looks in the street. Since its introduction into the market in 2009, more than 4,000 of these shelters have

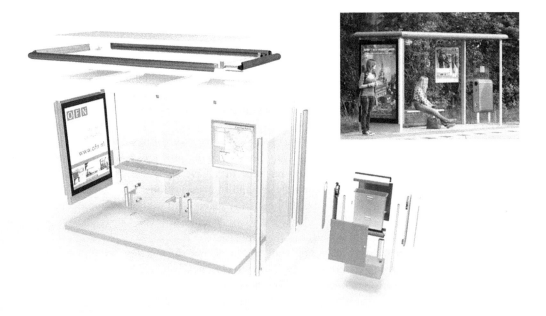

**FIGURE 2.1**

Exploded view of the "De Gouwe" bus shelter. (Inset: the "De Gouwe" in real life.) (For color version of this figure, the reader is referred to the online version of this chapter.)

*(Images courtesy of Fabrique Public Design, Delft, NL)*

been installed in the Benelux countries, in various versions and color schemes. Manufacturing this shelter involves a variety of processes: aluminum extrusion (support poles, roof frame, various smaller parts), metal casting (various parts of the bench and garbage bin), concrete casting (base plate, with inserts for easy assembly of the other components), and sheet metal forming (roof plates, various parts of bench and bin). It also involves a fair number of standard components that are bought in, such as glass panels and lighting elements. Assembly relies on screws, bolts, and other mechanical fasteners; welding (for the roof frame profiles); and clamping. Finishing involves powder coating (for sheet metal parts, certain extrusions) and anodizing (all other extrusions). In total, the De Gouwe shelter contains 729 parts, 168 of which are uniquely designed and manufactured for this shelter—that is, *product-specific parts* (of which 12 are custom-made fasteners). In total, it contains 575 fasteners. Some noteworthy details are as follows:

- All metal castings are made from aluminum using the sand casting method. This requires a modest investment for the casting molds. The method offers a large form freedom and is well suited for the relatively modest production volumes involved (Chapter 3). The castings are cleaned after manufacture and receive a double layer of powder coating (Chapter 13).
- Like the castings, all extrusions are also made specifically for this shelter. This requires some investment for the extrusion dies, but the benefit is that extrusion allows integration of all kinds of assembly functions into the extruded profile (Chapter 5); Figure 2.2 provides an example. The extrusions are either anodized or powder coated to get the right "look and feel" as well as corrosion resistance (Chapter 13).
- The sheet metal roof panels are made from galvanized low carbon steel. They are made by first cutting, or "blanking," the input sheet material into the right cutout shape, then bending its four edges, and finally powder coating. All of this can be done using universal equipment, so no investments are needed for these parts (Chapter 4).

The shelter involves three major sub-assemblies: the advertising box, the bench, and the garbage bin; the latter is shown in Figure 2.3. Note how it combines castings, extrusions, and sheet metal parts. The side panels are decorative only and, like the shelter's roof frame, can be powder coated in any color as desired. Such combinations of processes are in fact common, with complex parts being cast in metal or molded in plastic, less complex profiles being extruded or roll formed, and simpler parts being made by sheet metal forming or plastic thermoforming.

**FIGURE 2.2**

Cross section of extruded rear-right side profile, with integrated glass panes and fasteners.

*(Image courtesy of Fabrique Public Design, Delft, NL)*

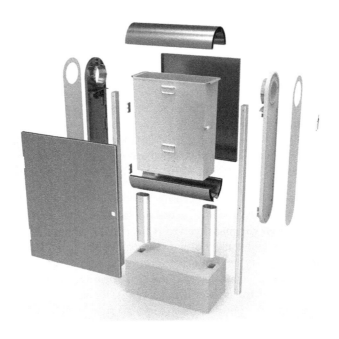

**FIGURE 2.3**

Exploded view of garbage bin, with cast concrete foot.

*(Image courtesy of Fabrique Public Design, Delft, NL)*

As a bus shelter, the De Gouwe is very successful, thanks to a number of key design decisions. First, it has a *modular design*, which allows a range of shapes, sizes, and color schemes with the same set of basic components. This modularity also allows fast assembly indoors, with a total assembly time of under four hours per shelter by a crew of one. Once assembled, the shelter is ready for placement in the street. Connecting it to the power network can be time consuming, but with an optional PV module and battery to provide electricity for lighting, the shelter becomes fully self-sufficient and can be placed in under 10 minutes. Furthermore, its minimalist design is not only aesthetically pleasing but minimizes the accumulation of street dirt and facilitates maintenance, which in turn keeps operating costs low. Of course, it is not vandal-proof, but cracked windows and other damage can be quickly and efficiently repaired. In fact, with the revenue from the advertisements in its poster frame, the shelter can actually bring in money. Durability helps, too: the design life span is 20 years, thanks to corrosion- and wear-resistant materials. Finally, costs have been reduced not only by the modular design but also by careful optimization of all parts, pushing the manufacturing technology to the limits without impairing quality. For instance, the wall thickness of all extrusions has been minimized, saving weight and, hence, costs.

Of course, despite its success, De Gouwe represents only one approach out of many possibilities, and several contemporary alternatives can be found, such as using steel for the main structure instead of extruded aluminum, or using a wooden bench instead of a metal one. The perfect bus shelter simply does not exist. But De Gouwe illustrates very well how various manufacturing processes can be successfully combined. Take a good look at a bus shelter, next time you find yourself in one, and try to see how it was made and which materials and processes were used!

## 2.3 INDOOR DESIGN: DOMESTIC EXTRACTION HOOD

In most domestic kitchens, you will find an extraction hood mounted on the wall above the stove, placed there to remove fumes generated during cooking. These hoods are usually made of steel and glass, and consequently they are mostly shaped like a flat box: the materials in question do not lend themselves well to more complex shapes, unless used for products made in large numbers, such as cars (more on this in Chapter 4). Installing these hoods is often a challenge, as they are heavy and usually require a tube connection through the kitchen wall or roof to the outside environment. Furthermore, changing the filter requires some time and skill and is therefore often done too late, causing smells to linger in the kitchen.

Working for the German kitchen appliances manufacturer Gutmann, the Dutch design bureau Van Berlo came up with an innovative solution. The hood design, shown in Figure 2.4, uses a fan-driven cyclone process to clean the air. Basically, dirty air rising from the stove is sucked into the hood and swirled around at high speed, flinging all vapors and fume particles against the collector vanes. The air is cleaned so well that it can be recirculated into the kitchen, so no connection to the outside air is needed (recirculation also saves energy, as no hot air is blown outside in winter). Thanks to this simpler design, a low overall weight of just under 10 kg, and several smart assembly solutions integrated into the design, the new hood can be installed in well under 10 minutes. The collector itself is easy to remove and can be cleaned quickly, either by hand or by putting it into a regular dishwasher. The product's overall shape is innovative as well: gently rounded, with a smooth, high-tech look and feel, without a single visible screw. Finally, as for price, the *L'Original* extraction hood, as it has come to be named, can easily compete against models from the leading brands in the kitchen appliance market.

**FIGURE 2.4**

Exploded view of *L'Original* extraction hood (left), and finished product (right).

*(Image courtesy of Van Berlo, Delft, NL)*

This striking and innovative design is possible thanks to a process that plays a vital role in the manufacture of countless products: *plastic injection molding*. In this process, further detailed in Chapter 8, molten plastic is forced into a metal mold at high pressure and rapidly cooled to shape the final part. The investment for the molds is considerable and the process is therefore only suited for larger production volumes (e.g., at least 10,000/year), but on the positive side, the freedom in forms, details, and materials is huge. This allows double-curved shapes to be made with optimized strength and stiffness, at low weight. Injection molding also enables the designer to integrate all kinds of functionality into parts, such as snap-fit joints and cable guides to allow fast assembly, thereby saving costs. Painting or finishing is not necessary; indeed, the process enables the manufacturer to quickly change the part color by selecting a differently colored plastic. In total, the *L'Original* hood contains 124 individual parts, including 59 fasteners (screws, clips, etc.) and 22 product-specific parts, 19 of which are made by injection molding. Some noteworthy manufacturing details include the following:

- For the overall product architecture, Van Berlo separated the structural parts, which carry all forces and act as an assembly base for all other components, from the non-structural, external covers. This allowed the company to select glass-filled nylon (PA66) for the structure, which is strong and stiff but not very attractive to look at, and an unfilled, good-looking ABS for the covers.
- The joints between the front and side covers were not made with conventional snap-fits, as these inevitably give rise to unsightly "sink marks" on the surface (see Chapter 8 for details). Also, as the higher temperatures above a stove may lead to material relaxation, loosening of snap-fits was a potential risk. Instead, a solution was found using two tiny aluminum extrusions, which simply slide over two rows of hooks on the two parts to be joined, as shown in Figure 2.5.
- Safety considerations demand that the 44 W motor and fan cannot be accessed while the motor is running. This was made possible by timer actuators that release the bottom mesh plate only when the motor has stopped turning. Also, trip switches have been integrated into the structure precisely where the mesh plate is clicked into place. Because there is no plate and no current, safety is assured (Figure 2.6).

Development of the *L'Original* extraction hood took roughly 1.5 years from first idea to market introduction at the Living Kitchen industry fair in Cologne, Germany, in January 2013. Up to that point, a significant development budget had been invested to make this product a reality. And with good reason: the hood immediately attracted much attention thanks to its striking design, ease of installation, improved usability and performance, low power consumption, and high filter efficiency. Currently it looks likely that the target production figures of some 50,000 units per year for a five-year production run will be easily reached.

As with the bus shelter, there is no perfect design for an extraction hood, and the *L'Original* is no exception. Market requirements are simply too varied to allow a single product to meet all demands, for all occasions. But the new hood does represent a radical departure from the traditional shape and operation of such products (and of manufacture: consequently, the steel-oriented Gutmann company also needed some time to adapt to the new materials and processes). This begs the question: which other household products that today are still steel-dominated could literally change shape using plastic injection molding? Of course, many products—vacuum cleaners, steam irons, hair dryers—have already made this switch, but not all of them. Take a good look around your house one day, and consider how manufacturing has not only shaped the way things look, but also how they can one day be transformed.

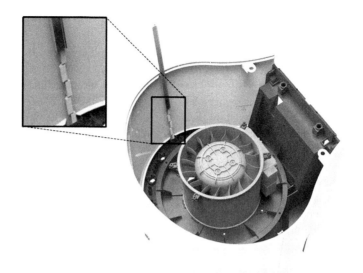

**FIGURE 2.5**

Detail of fairing joint with extruded rail (top view).

*(Image courtesy of Van Berlo, Delft, NL)*

**FIGURE 2.6**

Detail of mesh plate joint with integrated trip switch (bottom view).

*(Image courtesy of Van Berlo, Delft, NL)*

## 2.4 **PRODUCT DISASSEMBLY STUDIES**

In Exercise 2.1 you were asked to select a product that is ready for recycling or disposal. We now suggest that you disassemble this product, study its parts and components, and complete the exercises that follow. This is not only the ideal preparation for working your way through this book, but it is also an important part of what professional designers do—after all, why design everything new from the ground up when you can learn from existing products, thereby saving valuable development time? (Conducting detailed part and material audits for a product is also a key part of life cycle analysis to assess the environmental impact of products; disassembly studies, as shown in Chapter 14, also play a role in recycling.) Don't be fussy: since the product does not need to be re-assembled afterward, it is fine to cut through parts of your product to gain access to underlying components, use a power drill to destroy screws or rivets, and so on. But whatever you do, always think about your safety and that of any bystanders first: for instance, never disassemble a product that is plugged in or could be inadvertently powered up. In general, we urge you: *be sensible, be safe!*

---

**EXERCISE 2.2**

How many individual parts did you encounter, including all screws, clips, and other fasteners? Is this more than you expected? Did you sometimes find ambiguity in defining what was a single part and what was actually a collection of parts joined together (i.e., a sub-assembly)?

---

**EXERCISE 2.3**

Which parts are product-specific, meaning that they were specifically designed for this product? Conversely, which parts are bought in as standard components and are not product-specific?

---

For this next exercise, note that product-specific parts can be sub-divided into two categories. The first of these consists of all parts that require dedicated investments, such as the castings and extrusions in the bus shelter discussed in Section 2.2 The second category comprises all parts that require no investments but that do require specific operations to give them the proper shape, dimensions, and look and feel, such as the shelter's sheet metal roof plates. Non-product-specific (or standard) parts are items such as screws and other fasteners, but also parts such as bearings, bushings, electric motors, LEDs, and other mass-manufactured items. In nearly all everyday products, these standard parts outnumber the product-specific parts by far, but the product-specific parts are the ones that literally give shape to the product.

---

**EXERCISE 2.4**

Select three product-specific parts. Which materials are used? How can you tell?

As for materials, the first and obvious distinction to make is among the four main categories of metals, plastics, ceramics/glasses, and hybrids. Next, we can distinguish among different kinds— for metals: plain carbon steel, stainless steel, wrought or cast aluminum, and so on; for plastics: polypropylene (PP), acrylonitrile butadiene styrene (ABS), polyamides (PA), polycarbonate (PC), and the like. (Note that especially with plastics, you will often encounter trade names, such as "nylon" for PA or "Lexan" for PC.) More detailed identification of specific alloys, or plastic grades, will usually require specialist knowledge and equipment. But simple experiments can help distinguish between materials that can look similar, such as zinc and aluminum—for example, you could try measuring the density of a component by using a principle known since Archimedes' times.

To identify metals, it is essential to know that plain carbon steel—the most commonly used metal, thanks to its low cost, good strength, and excellent formability as sheet metal—is nearly always coated for corrosion resistance, with paint, zinc-, chrome- or tin-plate, or a polymer coating. In other words, what may at first sight look like stainless steel is often the considerably cheaper chrome-plated carbon steel. (How can you tell the difference? Austenitic stainless steel is not magnetic, whereas carbon steel is—a huge benefit in sorting mixed metal scrap for recycling.) To identify plastics, a good hint is that in contemporary products, the type of plastic is often designated on the part (e.g., "PP" for polypropylene or "PC" for polycarbonate) or, alternatively, via a number code within a triangular recycling symbol (1 for PET, 2 for HDPE, and so on). (It is not always this easy: in one product disassembly case study, the designation "AS" on a plastic part turned out to be short for "anti-static," not for some kind of polymer.) And a final tip: if you have access to the Cambridge Engineering Selector (CES) EduPack, note that this database also includes typical applications of each material it contains, and these applications are searchable.

---

**EXERCISE 2.5**

For the same three parts, which manufacturing principle (e.g., casting of metals, injection molding of plastics) has been used? What design features or evidence on the product suggest your choice of process?

---

Unless you have considerable prior knowledge of manufacturing, this last question is, of course, not easy to answer. However, it will become steadily clearer as you work through the book; you will come to recognize why a certain process has been chosen for a certain part and how this choice has influenced both the final material choice and the part's precise shape.

## 2.5 **Further Reading**

Another bus shelter design by Fabrique, made for the city of Amsterdam and JCDecaux, was featured in a Discovery Channel episode of *How do they do it?* (first aired in February 2013). For a video of how the *L'Original* extraction hood was designed, see www.youtube.com/watch?v=39wRbgY2I70.

For more examples of exploring the design of everyday products, including exploded views, we recommend the following references:

Ashby, M.F., 2011. Material Selection in Mechanical Design, fourth ed. Butterworth-Heinemann, Oxford, UK. (*An accessible design-led text covering the selection methodology behind the CES software, but starting with an interesting take on the way that products have evolved to exploit new materials.*)

Industrial Designers Society of America, 2001. Design Secrets: Products—50 Real-Life Projects Uncovered. Rockport Publishers, Minneapolis, United States. (*An excellent book showing how products are designed, with considerable attention to materials and manufacture.*)

www.ifixit.com. (*This do-it-yourself repair site routinely takes apart the latest products, with a certain focus on electronics. Highly recommended!*)

www.toddmclellan.com. (*A one-man venture, but what an effort: this Canadian photographer turns product disassembly into an art form.*)

# Shape Casting of Metals

3

## CHAPTER OUTLINE

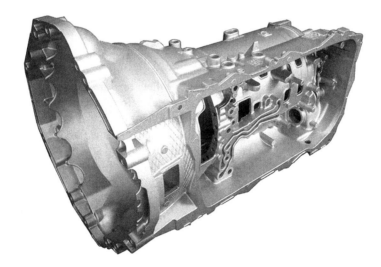

This housing for a seven-speed automotive gearbox, made by high pressure die casting of aluminum, represents well what casting of metals stands for: high strength, large design freedom. Image courtesy Brabant Alucast.

## 3.1 **INTRODUCTION**

Shape casting of metals is one of the key manufacturing processes. This importance is mainly due to the large form freedom that this technology offers: in just one step, it allows the manufacture of complex shapes that would otherwise need to be assembled from separate parts. In addition, the products—or, more to the point, the *castings*—possess the attractive properties we expect from metals, such as high strength and stiffness. Castings also retain their strength at elevated temperatures, and you can therefore expect to find castings in a kitchen stove, a domestic iron, and the engine block and cylinders of a car. Even jet engine turbine blades are cast (from special nickel alloys). A third reason for the popularity of castings is the distinct character inherent to them, with flowing, organic shapes and distinct textures. Many *art nouveau* objects, such as the cast iron arches of Parisian metro stations, as well as countless cast bronze statues, testify of this character and the artistic potential of the process.

The oldest metal castings date back to around 5000 BCE, and large-scale iron casting helped to launch the Industrial Revolution, but it would be a mistake to think that the process is outdated. Relatively young metals, in particular aluminum, zinc, and magnesium, have given it new life, and even the seemingly old-fashioned casting of iron has been markedly improved, allowing for thinner, lighter, and stronger components. However, despite these developments, the three benefits of form freedom, strength, and character still come at a price, especially if we need to manufacture high volumes and therefore require short cycle times. Stringent quality demands, in particular narrow tolerances, good material properties, or near-perfect surface finish will also increase cost. It is not easy to understand how all of these variables correlate. The fact that industrial practice uses an increasingly wide range of different casting processes— or *manufacturing methods*—does not make things easier. Therefore, we need to take one step back and consider what really happens at the material level when we make a casting. This will be the theme of this chapter. A small number of elementary physical considerations give a surprisingly strong grasp of what the various trade-offs among function, quality, and cost look like for shape casting of metals.

Casting is also used to make the ingots, slabs, or billets that are the starting points for making "wrought" products (i.e., as the input for solid-state metal deformation processes). Hence, many of the fundamentals of casting are relevant here too, but without the issues of shape complexity. Many consequences of this first casting step—surface finish, porosity, microstructure—carry forward in important ways to influence the subsequent rolling, extrusion, and so on. We'll see some examples in later chapters. For this chapter, the focus is on *shape* casting of products.

---

**EXERCISE 3.1**

Look around your house and find one or more examples of cast parts or products. How important do you think form freedom, strength, and cost were in choosing casting for these products?

---

## 3.2 **FILLING THE MOLD**

In shape casting, a certain amount of liquid metal is poured into a cavity that defines the negative shape of the product: the *mold*. This can be done in a variety of ways and using any amount of pressure. Figure 3.1 shows a schematic of how this can be done if we simply pour the metal in. Observe that

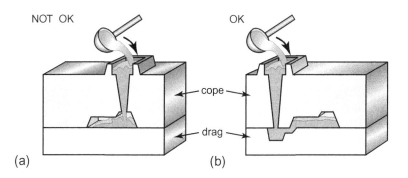

**FIGURE 3.1**

Mold filling: (a) top-down; (b) bottom-up (air outlet channel, the "riser," not shown.) (For color version of this figure, the reader is referred to the online version of this chapter.)

if we pour "top down," the molten metal will splash, creating a lot of free surface where the metal will immediately oxidize. Particularly in the light metals aluminum and magnesium—both are commonly used for castings—this leads to loss of material properties, especially ductility. If we pour "bottom up," we do not have this disadvantage. As an added benefit, the air inside the mold, as well as any gases and vapors that are released during casting, can evacuate more easily, without causing air bubbles that may become entrapped in the metal and that also deteriorate properties. As Section 3.7 will show, not all methods leave much choice in the matter but, if possible, casters always prefer to fill molds from the bottom up.

If the molten metal has a low viscosity, it fills every detail of a complex mold well, without the need to apply extra pressure. In this respect, there is a key difference between steel (i.e., iron with less than 2% carbon, and typically just 0.2% to 0.6%) and cast iron (i.e., iron with 2% to 4% carbon content), or between aluminum alloys with small amounts of alloying elements and those with 6% to 12% silicon: in both cases, adding the right amount of a key element significantly lowers the viscosity of the molten metal. A major practical benefit is that the melting temperature is also reduced. Plain carbon steel, for instance, melts at around 1,500°C, but cast iron at some 1,200°C (there is even a third benefit, which we will explore in Section 3.3). This explains why steel is almost never used for shape casting—although it is possible and, indeed, very common to cast it into slabs or ingots. Viscosity and melting point reach a minimum at the *eutectic composition*. Alloying metals to be at, or near, this composition is nearly universally applied in casting of metals, increasing what the casting world refers to as the "fluidity" of the molten metal, and reducing the processing temperature (giving lower energy costs and faster throughput). Note that only pure metals or eutectic alloys melt (and freeze) at a single temperature; at other compositions, we see a melting range.

**EXERCISE 3.2**

Look up the compositions of several zinc and magnesium casting alloys (at least two of each). Also, find a typical composition of brass. What are the key alloying elements? How do the melting temperatures of all of these alloys compare to those of the pure metals?

---

**EXERCISE 3.3**

Do you think that "18/8" stainless steel (AISI 304) can be cast as well as cast iron? In fact, do you think it can be shape cast at all, for practical purposes?

---

So for ease of castability, castings generally have much higher alloy content than the typical compositions of wrought alloys, but this does have a major drawback: it comes at the expense of mechanical properties. Even if the castings are free of defects—which they rarely are—the toughness, ductility, and fatigue strength in particular are reduced. (Later chapters will show how we enhance and control these properties via wrought processing and heat treatment instead.) Cast iron represents a special case where the stiffness too is significantly lowered, as compared to low carbon steel. So there is a major trade-off to be made, optimizing between properties and castability.

Looking deeper into mold filling, we see that it is not just a question of bottom-up versus top-down filling: we also have to control the filling speed, depending on the length that the liquid metal must travel (i.e., the flow length). This has its basis in the difference between laminar and turbulent flows, shown in Figure 3.2. The former has the minimum free surface area on which the molten metal can oxidize, and consequently the least loss of material properties (again, this is especially relevant for the light metals). Flows start off as laminar and, after a certain critical flow length, become turbulent, with much more free surface and hence more oxide formation. Smoke curling upward often reveals this transition well; it also applies to flowing liquids.

The transition takes place at a certain Reynolds number ($Re$), which is a dimensionless number defined as $Re = \rho v l / \mu$. Here, $\rho$ is the density in kg/m$^3$, $v$ the flow speed in m/s, $l$ the flow length in m, and $\mu$ the viscosity in Pa.s $=$ kg/(m.s). The transition value of $Re$ varies somewhat between metals, but to give a first idea: in aluminum casting, flow speeds below 0.5 m/s are generally "safe," whereas above 1 m/s, turbulence almost certainly occurs. Note that this depends not only on flow length but also on the mold geometry and surface roughness.

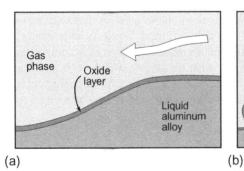
(a)

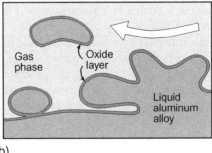
(b)

**FIGURE 3.2**

(a) Laminar flow, and (b) turbulent flow, with free surface and oxide layer. (For color version of this figure, the reader is referred to the online version of this chapter.)

Are high viscosities beneficial (i.e., do they postpone the transition to turbulence to longer flow lengths or higher filling speeds) or detrimental for mold filling?

So low filling speeds are good for quality, but they also mean that filling takes more time, lowering productivity. Again, we see a trade-off. If we pour a thin-walled product slowly, we run the risk that the metal solidifies before the mold is fully filled: a fatal casting defect known as "cold running." Filling the mold via multiple openings (known as "gates") can prevent this, as the length that the metal then has to flow to fill the entire product is reduced. However, each gate requires finishing afterward, so this incurs a cost. Another option is to increase the thickness, but that adds weight and, again, cost. In practice, mold filling is done either slowly, aided only by gravity or low pressures, or quickly, using high pressures, accepting that there will be consequences for the material properties.

Beneath its white plastic shell, the Apple G4 iMac had a hemispherical support frame (Figure 3.3) weighing 2 kg and cast in Zamak, a common zinc alloy. With part thicknesses as low as 3 mm, the filling time had to be less than 0.2 s to prevent cold running. Do you think this involved laminar or turbulent filling? Explain!

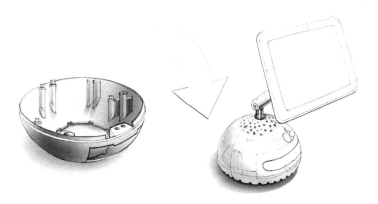

**FIGURE 3.3**

iMac G4 (right) and support frame (left). (For color version of this figure, the reader is referred to the online version of this chapter.)

---

**EXERCISE 3.6**

Look closely at the products you selected for Exercise 3.1. Can you see where the gate (or gates) was placed to fill the molds? Can you also spot any air vents? Which metals and alloys have been used in these castings, and how can you tell?

---

A quick note on terminology: high filling pressures require the mold to be made of metal (tool steels, usually). In the world of metal casting, a metal mold is referred to as a *die*. Strangely enough, in injection molding of thermoplastics (see Chapter 8), the metal dies used there are again called *molds*. This simple case illustrates how terminology does not always carry over among different principles, serving as another reason to study each principle in some detail.

## 3.3 THE SOLIDIFICATION PROCESS

After mold filling, the molten metal has to solidify. In physical terms, it is this solidification process that primarily governs casting of metals (i.e., the phase transition from liquid to solid). In this transition, the change in the metal's volume with temperature is typically as shown in Figure 3.4.

We start with liquid metal in the top-right corner of the graph. As the metal cools down, it contracts at a rate that is typically in the range 0.01% to 0.02%/K. This continues until the onset of solidification at temperature $T_{liquidus}$. Most metal casting alloys are non-eutectic and have a freezing range, only becoming fully solid at the lower temperature $T_{solidus}$. Note from the graph that solidification is accompanied by a more rapid decrease in volume. This "shrinkage" is due to the atoms being neatly ordered in

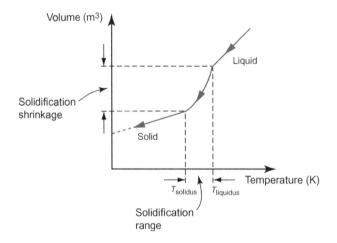

**FIGURE 3.4**

Volume against temperature during solidification. (For color version of this figure, the reader is referred to the online version of this chapter.)

**Table 3.1** Key Material Data for Several Casting Alloys

| Material Property | Aluminum (Al$_{12}$Si) | Gray Cast Iron (Fe$_3$C$_2$Si) | Zinc (Zn$_4$Al) | Bronze (Cu$_{10}$Sn) | Magnesium (Mg$_9$Al) |
|---|---|---|---|---|---|
| Solidification range liquid → solid, °C | 584-572 | 1240-1130 | 387-381 | 1000-854 | 596-468 |
| Heat of solidification (latent heat), kJ/kg | 389 | 270 | 112 | 230 | 373 |
| Typical volumetric shrinkage, % | 3.8 | 1[†] | 6.5 | 4.5 | 4.2 |
| Specific heat, solid phase, J/(kg.K) | 963 | 500 | 420 | 384 | 1090 |
| CLTE,* m/(m.K) | $21 \times 10^{-6}$ | $13 \times 10^{-6}$ | $27 \times 10^{-6}$ | $18 \times 10^{-6}$ | $26 \times 10^{-6}$ |

*CLTE is coefficient of linear thermal expansion. To convert to volume thermal expansion (or contraction) per kelvin, multiply by 3: for example, cooling Al from 555°C to RT gives 3.4% volume contraction.
[†]Gray cast iron shows shrinkage values between −2.5 and +1.6, depending on the C content.

a crystalline structure in the solid state, occupying less volume than in the disordered liquid state. Once it has fully solidified, the metal can be cooled down further to room temperature (RT for short), contracting again. Notice how the graph's slope is now less than it was in the liquid state. The slope is characterized by the thermal expansion coefficient, defined as the linear strain per kelvin change in temperature. The numerical values of all of these properties are material dependent; Table 3.1 presents some key data for five important casting alloys. Observe the low shrinkage of gray cast iron (due to the formation of low-density graphite in the solid state, compensating for the iron contraction). Depending on the carbon content, it can be zero, or even expand on solidification (negative shrinkage value). Low shrinkage is the third benefit of this material compared to cast steel alluded to in the preceding section.

---

**EXERCISE 3.7**

How can we deal with the contraction between $T_{solidus}$ and $RT$ in the design or casting of products? How can we deal with shrinkage?

---

**EXERCISE 3.8**

Why do different alloys have different shrinkage? Hint: consider the possible differences in crystal structures, and also consider what happens to certain alloying elements before and after solidification.

Our next question is, how much heat (in joules) must be dissipated during the casting process? Just like the volume loss, this consists of three parts. First, we must dissipate heat during the liquid phase i.e., to get from our starting temperature, also known as the pouring temperature, to the solidification temperature. The material property involved here is the specific heat capacity, with units of J/(kg.K). Second, we must dissipate the latent heat of solidification, which is released progressively while going from $T_{liquidus}$ to $T_{solidus}$, with units of J/kg. Finally, the specific heat capacity of the solid between solidification $T_{liquidus}$ and RT must be dissipated, again in J/(kg.K). All three properties depend on the alloy—again, refer to Table 3.1 for data. In practice we may not need to wait until the casting has cooled down to RT inside the mold: sometimes we can "de-mold" earlier and start the next casting cycle sooner, increasing productivity.

---

**EXERCISE 3.9**

For each alloy in Table 3.1, determine the amount of heat that must be dissipated to cool 1 kg of this metal from the bottom of its solidification range ($T_{solidus}$) down to RT. Compare your answers to the latent heat of solidification of the various metals. Which is bigger, and by what factor?

---

**EXERCISE 3.10**

In practice we can de-mold when the casting is cool enough that it will have sufficient hot strength. Assuming this occurs at a temperature of 0.5 $T_{solidus}$ (in kelvin), revise the analysis made in the previous exercise.

---

Recall that the mold is filled through one or more gates. In addition, we will need exit gates to allow the air to be expelled. The inlet and outlet channels are referred to as "runners" and "risers," respectively. The material in these channels will also solidify and must be separated from the actual product. This is easiest if the gates are thin, so that we can simply break the material off at that point. If the gates are situated at non-visible locations on the casting, then no further finishing is needed, but if they are in plain sight of the user, these unsightly marks must be removed by machining. So gate placement has not only to do with efficient filling.

Even more importantly, we need to deal with shrinkage. One solution is to locate the runners and risers on the thicker sections of the product. As explained further later, these sections will be the last to solidify and will experience the greatest shrinkage. Shrinkage in the thinner sections is compensated by the inflow of extra liquid from the thicker sections to which they are attached. Similarly, properly placed feeders can supply the thicker sections (as their name implies), compensating for shrinkage in the casting. For this to work, the runners and risers must solidify last, which means they need to be relatively large (for reasons explained later). Notice, therefore, that the "pour weight" of the casting (i.e., the weight of liquid metal used) is always greater than the weight of the casting itself—easily 40% extra or more. Even though the metal in the feeding system is recycled within the foundry, energy costs scale with the pour weight, so minimizing the surplus is part of the art of mold design (known as "methoding").

---

**EXERCISE 3.11**

Glass is very different from metals but can nevertheless be cast quite well. What do you think would be the shrinkage for glass? And which property of molten glass allows a skilled glassblower to mold and shape the material on the end of a hollow rod in the open air, whereas this is impossible for metals?

---

## 3.4  DIGGING DEEPER: THE SOLIDIFICATION TIME

The production rate and economics of a casting process depend on how long it takes to get the heat out, such that the cast metal part is cool and strong enough to be handled. We therefore now ask this question: how long does solidification and cooling take, and on which variables does this depend? As it turns out, we can make a surprisingly simple model for the solidification time, at least for idealized situations. First, notice that the amount of heat we must dissipate depends on the volume $V$ of the casting. Second, this heat can only be dissipated through the casting's surface area $A$. Simple heat flow considerations tell us that the solidification time scales with $V$ and $A$ according to *Chvorinov's rule*, first published in 1940: $t_{solidification} = C\,(V/A)^2$. Here, $C$ is the Chvorinov coefficient (units: s/cm$^2$), which depends on both the casting metal and the mold material—typical values are shown in Table 3.2. The rule gives valuable insight into the design of castings—after all, the product's volume and surface area are under the direct control of the designer. Given certain choices for the casting metal and mold material, we can minimize the solidification time by minimizing the *thermal modulus* (i.e., the ratio of volume to surface area, $V/A$).

---

**EXERCISE 3.12**

On skateboards, the wheels are mounted onto the board by means of a "truck," which consists of a "hanger" and a "base plate." Both are usually aluminum castings (Figure 3.5). Using Chvorinov's rule, estimate the solidification times of these parts by modeling them as simple geometrical shapes. Which takes longest?

---

**Table 3.2** Chvorinov Constant $C$ for Various Metals and Molds, in s/cm$^2$

| Mold Material | Sand Mold | Steel Die |
|---|---|---|
| **Alloy** | | |
| Gray cast iron | 65–70 | 2–2.4* |
| Zinc | 100–180 | 2–6 |
| Aluminum | 90–120 | 1.6–2.4 |
| Bronze | 40–60 | 2.2–3.4 |
| Magnesium | 30–50 | 0.5 |

*Ranges depend on pouring temperature as well as mold temperature.*
*In practice, this combination of casting metal and mold material is extremely rare.*
*Source: Mark Jolly, Cranfield University.*

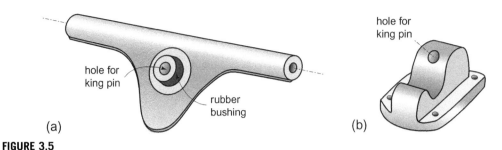

**FIGURE 3.5**

Skateboard truck parts: (a) hanger; (b) base plate.

---

**EXERCISE 3.13**

In two-shift production, we typically have around 3,000 hours per year of effective production time. What then is the maximum time to produce one part (i.e., the "cycle time") if we want to produce 60,000 skateboard trucks per year? Will the solidification time you estimated in Exercise 3.12 allow this volume?

---

Chvorinov's rule can be applied quite well to relatively massive castings, for which the volume of material involved essentially determines the speed of solidification. For hollow castings, such as an engine block or a cooking pot, it becomes more difficult to use the rule effectively, because you cannot simply assume that the mold material on the "inside" of such castings absorbs heat as easily as that on the outside (in other words, the effective surface area $A$ will be less than you might think). So, in such cases, some caution is called for in applying the rule.

---

**EXERCISE 3.14**

Estimate the solidification time for a brass church bell measuring 20 cm across and 30 cm high, with a 15-mm wall thickness (model as a hollow cylinder, open at one end). Do this for a worst-case scenario (i.e., zero internal surface area) as well as for a best-case scenario.

---

For thin-walled castings (i.e., less than ∼4 mm thick), Chvorinov's rule strongly underestimates the solidification time, no matter how careful you are. This is because for such castings, the rate of heat transfer through the contact surface between mold (or die, if made of metal) and casting becomes dominant. Consider that during solidification, the mold heats up and expands, while the casting cools down and contracts. An air gap is then formed between mold and casting, acting as an efficient insulator and strongly resisting the flow of heat. The result is that for thin-walled castings, solidification time is no longer dependent on $(V/A)^2$ but simply on $V/A$. Note that once more, minimizing $V/A$ is the key to productivity.

---

**EXERCISE 3.15**

For a thin, flat plate with thickness $d$, the thermal modulus can be approximated by $d/2$. Explain why.

---

**EXERCISE 3.16**

Estimate the solidification times for your various sample products (see Exercise 3.1).

---

In rounding off this section, we should point out that the thermal history of a casting has a major influence on the quality corner of the manufacturing triangle, not just the costs. The cooling history determines the elastic-plastic strains that may be imposed on the casting while it cools down, leading to shape distortion and residual stress. Furthermore, cooling rate is central in determining the solidification *microstructure* and the material properties of the casting (particularly strength and ductility). These issues are discussed in the next two sections.

## 3.5 CASTING DEFECTS: POROSITY, INTERNAL STRESSES, AND DISTORTION

Of the many quality issues that we have to live with in casting of metals, three of the main ones are porosity formation, and the occurrence of internal stresses and distortion. Here these are covered in this order—as we shall see, the last two are closely linked.

Porosity comes from two sources. First, it is the result of shrinkage that is not fully compensated by the inflow of extra material. Imagine that we try to fill a thick section of a casting through a thinner one. Once the thin section solidifies, it blocks the inflow of material to the thick section, which at that point in time consists of an already solidified skin surrounding a still-liquid core. Inevitably, cavities are then formed in the thick section to make up for the loss of volume during solidification. The same effect occurs if the runners and risers are unable to feed liquid metal into the casting. Such "shrinkage porosity" can be minimized by filling from thick to thin and by careful design of the feeder system (though using many feeders will incur considerable extra costs, as more points will require machining and finishing afterwards, and more production scrap must be recycled). The second source of porosity can never be eliminated completely. Hot liquid metal dissolves gases much more easily than solid metal. So during solidification, gases such as oxygen and carbon dioxide are released. Some bubbles will escape if they can find a route upwards through the still-liquid metal, but many become trapped within the solid. It is the reason that water is transparent, but ice is not. The trick is to make sure that the gas gets dispersed into many tiny bubbles, giving "microporosity" with only a small impact on the material properties, rather than allowing it to collect in one place in the last part to solidify ("macroporosity"). This is closely related to the way that the solid constituents are distributed in a casting—we'll come back to this discussion in Section 3.6.

---

**EXERCISE 3.17**

Which of the alloys in Table 3.1 is probably the most susceptible to porosity formation? Why?

---

Next, consider internal stresses and distortion. The preceding section shows that the solidification time depends strongly on the shape of the casting: the higher its thermal modulus $V/A$, the longer that solidification will take. The same goes for the time that must elapse between full solidification and de-molding (i.e., the cooling time). We can apply these insights also to different sections of a single casting: thin sections will solidify and cool quicker than thicker ones. The different cooling rates will lead to temperature gradients within the casting.

---

**EXERCISE 3.18**

Again, consider the G4 iMac frame. This casting has an average wall thickness of 3 mm, but it has local sections that measure up to 15 mm thick (see Figure 3.3). How do the cooling times of these thicker sections compare to those of the thinner wall?

---

As Table 3.1 shows, common casting alloy metals expand when heated and contract when cooled. A simple formula describing this effect is $\varepsilon_{th} = CLTE\ \Delta T$, with $\varepsilon_{th}$ being the thermal strain, $CLTE$ being the coefficient of linear thermal expansion (commonly given in $10^{-6}$ m/(m.K)), and $\Delta T$ being the temperature change. Therefore, differences in temperature within one casting will cause differences in contraction. According to Hooke's law ($\sigma = E\ \varepsilon$), these differences then cause stresses. Where these exceed the material's yield stress, we will end up with either permanent deformation (if the casting is free to deform) or internal stresses (if it is not free to deform, for instance, because its own shape and the surrounding mold restrict deformation: in this respect, sand molds and steel dies behave quite differently). If the stresses get even higher and the metal is not very tough, castings may actually crack and rupture! Sometimes, the internal stresses can balance out across a section, preventing any distortion. However, if such a casting is then later machined on one side, it will still warp, as the stress balance gets broken. Or if the component is loaded in service, the internal stresses are superimposed on the service loads—a particular issue in accelerating fatigue failure.

---

**EXERCISE 3.19**

Consider a cast iron manhole cover with ribs for increased stiffness. Suppose that during casting we get a certain temperature difference $\Delta T$ between the thicker and thinner sections of this product. How large does $\Delta T$ need to be to cause permanent deformation (typically at a strain of order 0.1%)? Is this difference likely to occur during casting?

---

In practice, we must rely on computer simulations to predict the internal stresses and the risk of deformation or rupture. After all, casting is usually used to produce complex shapes that are impractical or even impossible to realize with other manufacturing principles, and such shapes do not lend themselves to "back of the envelope" calculations. Also, the yield strength and Young's modulus of materials are not constant but decrease with increasing temperature: only a computer can then keep track of what is going on. Still, even simple calculations can give a grasp of the problems at hand.

---

**EXERCISE 3.20**

Look closely at the parts you selected (see Exercise 3.1). Do you think that porosity or internal stresses or distortion are present? Explain. Hint: measure your products carefully and estimate the thermal moduli of the various sections. Another hint: cut right through the thickest section and see if you can see any porosity—polishing the section and using a microscope will help, but in cast aluminum, you can often see porosity with the naked eye.

---

Can these quality problems be avoided? With proper design and manufacturing, yes, and usually at low cost. The key elements of the solution are to choose the right alloys, to avoid strong changes in section thickness, and to give castings a flowing, organic shape, without sharp radii.

## 3.6 CAST MICROSTRUCTURE AND PROPERTIES

Casting metallurgy is a large subject with a long history, full of all sorts of alloy-specific details and "tricks" developed empirically over the years to improve cast properties. In the spirit of this book, we will only touch on a few key principles to give a flavor of the main issues. The books listed under the "Further Reading" section provide ample additional detail.

First, recall that metals have a crystalline *grain structure*. Solidification microstructures are *polycrystalline*—that is, they are made up of *crystal grains*, each with its own orientation of the atomic lattice. This is a result of the nucleation of many small volumes of solid, forming on the mold wall or within the melt, each growing until all of the liquid has been consumed. Where growing grains impinge on each other, a *grain boundary* is formed. Cast grain sizes are typically of order 0.05 to 5 mm. Grain structure has two main consequences: (1) key properties such as strength and toughness improve as grain size is reduced; (2) casting alloys containing many elements and the homogeneity of their distribution depends on the grain size—this *segregation* of alloying additions merits further explanation.

The alloying elements in a cast metal fall into four categories: (1) additions to improve the processability (e.g., silicon in aluminum to lower the melt viscosity and melting point); (2) additions to improve the properties, particularly the strength, by the formation of two-phase microstructures (same example, but now because the silicon forms as hard, fine-scale crystal regions dispersed within the aluminum grains); (3) impurities—unavoidable traces of elements from the original extraction process, or recycling, usually having a damaging effect on properties; (4) minor additions to minimize the detrimental effects of impurities or to modify and improve the microstructure.

To understand the detail of all this requires in-depth knowledge of phase diagrams and phase transformations, for which the reader is referred elsewhere (see "Further Reading"). But there is one central effect: in the liquid state, virtually everything dissolves readily into an atomic soup; but when solid, the elements separate out into multiple phases, each of which contains just a few elements. For example, the grains of a cast aluminum alloy are predominantly crystals of aluminum with a small amount of stuff left dissolved in it, but embedded within the grains are countless small crystals of pure silicon and various intermetallic compounds, plus impurities. As noted earlier, some of the impurities dissolved in the liquid are gases, but the solid rejects these gases during solidification, leading to porosity.

Segregation is the name given to the non-uniform distribution of the dissolved elements that results from solidification. It stems directly from the different solubility of elements in liquid and solid, over the freezing range of the alloy. The first solid to form is purer than the average composition, so the residual liquid is enriched in alloy elements and impurities as a result. The last part to solidify will always be at a grain boundary. This is the heart of the problem. If impurities or gas bubbles are concentrated at grain boundaries, there can be a serious loss of resistance to fracture, fatigue, or corrosion. The shape of the solidifying grains also matters. If the solid-liquid interface grows inward from the mold surface toward the center as a planar front with just a few grains, elements can segregate right across the casting—the middle of thick sections is the place to look for the worst porosity. But if we can form an equiaxed grain structure, the elements will be more uniformly spread throughout the casting. So the length scale of the inhomogeneity, and the severity of the segregation, depend on grain structure—what we want are fine equiaxed grains to minimize the problems.

So what determines the grain size and shape? Principally it is the *cooling rate* imposed by the process—notably depending on the type of mold used (and, hence, on the method of casting: see Section 3.7), the size and shape of the casting itself, and which alloy we are casting. So we have a glimpse here of a deep complexity—for a given casting alloy, a big casting will have poorer properties than a small one; and the same design cast in a sand mold or in a metal die will not yield the same properties either. The most challenging issues in design and manufacturing are these closely coupled problems, where parameters of the material, the process, and the design combine in complex ways to determine a key quality outcome.

---

**EXERCISE 3.21**

Estimate the number of grains across the thinnest section of the casting you identified in Exercise 3.1. Hint: about the cooling rate?

---

Casting metallurgists have come up with many chemical interventions to help control microstructures. *Inoculants* are high-melting-point powders added before pouring (e.g., $TiB_2$ added to aluminum casting alloys). The tiny solid particles promote "heterogeneous nucleation" of the first solid to form from the melt. Because each nucleus forms a grain, more nuclei mean smaller grains and less segregation. Impurities can sometimes be rendered harmless by adding something else to react with the impurity—for example, a particularly clever trick used to take oxygen out of harm's way in steel ingots is the addition of aluminum, forming tiny particles of (solid) aluminum oxide within the grains, rather than allowing the gas to segregate to the grain boundaries to form porosity. Finally, some second phases in common casting alloys are inherently brittle—such as silicon in aluminum alloys, or graphite in gray cast iron. These phases also tend to form as sharp flakes within the solid, making them behave like microcracks and giving poor fracture resistance. But tiny additions of just the right element can completely change the size and shape of the brittle phases, giving remarkable improvements in tensile strength—a trick known, for some strange reason, as "poisoning"—for example, a trace of sodium does wonders for aluminum casting alloys.

---

**EXERCISE 3.22**

Which element(s) can be used to ensure that the carbon in cast iron emerges as nodules ("nodular cast iron") instead of as flakes ("lamellar cast iron")? Look it up! And what is the key difference in material properties and, hence, applications?

---

**EXERCISE 3.23**

Look up a cast alloy composition in detail. What are the impurities? What elements are deliberately added, and why?

---

**EXERCISE 3.24**

Do you think inoculants will affect the cycle time?

Specialist knowledge is clearly needed here. The take-home message is to be aware that properties quoted for casting alloys are often for "ideal" conditions, and there may be significant knockdown factors. Especially for safety-critical components, you really need to know what you are doing in choosing alloy, process, and process conditions to hit the property targets for your particular design.

## 3.7 SHAPE CASTING METHODS

This section summarizes the strengths and weaknesses of the main manufacturing methods used for casting of metals, explaining them using the insights obtained in the previous sections. Shape casting methods are distinguished from one another by looking at (1) the manner of mold filling and (2) the mold material. Here we will discuss sand casting, gravity die casting, low-pressure die casting, high-pressure die casting, and investment casting.

### Sand casting

Silica sand, mixed with a suitable bonding agent, is by far the cheapest of all mold materials. Sand molds are relatively easy to make and the part-related investments of sand casting are low. Wooden, plastic, or metal "patterns" are used to form molds containing a negative of the shape of the casting: the mold cavity (Figure 3.6). Another plus for this method: it can deal with all common casting metals (notably cast iron), and a wide range of product sizes, from grams to tons per product. However, sand molds are relatively fragile and can only withstand low filling speeds and pressures; furthermore, they are poor conductors of heat, leading to long solidification times, coarser microstructures, and poorer properties. They also have a rough surface quality and comparatively poor tolerances.

A sand mold consists of two main parts: the lower and upper mold halves, called "drag" and "cope," respectively, by professional casters. The plane where these two halves meet is referred to as the *parting plane*, which is usually oriented horizontally. Relative to this plane, the casting must have a suitable draft angle. This is not only to allow easy de-molding (after all, the mold is used only once and could therefore also be destroyed instead), but also because the mold parts themselves must be made as well: so it must be possible to remove the patterns from the sand. However, separate sand cores, which

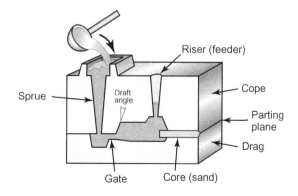

**FIGURE 3.6**

Sand casting schematic. (For color version of this figure, the reader is referred to the online version of this chapter.)

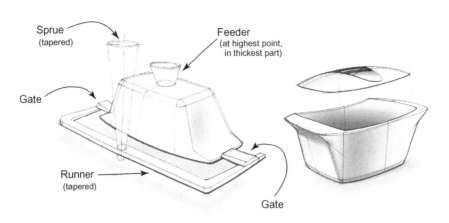

**FIGURE 3.7**

Sand casting of the "*Le Creuset*" cooking pot (left: upside-down product with casting system; right: finished product).

can be easily removed by vibration after solidification and cooling, allow the manufacture of hollow products (see Figure 3.6).

Usually, sand molds are filled through a downward channel called a "sprue" merging into one or more runners leading to the gate(s), as shown in Figure 3.7 for a true design classic, the cast iron *Le Creuset* cooking pot. The mold cavity is filled from bottom to top so that the air can easily escape and the liquid metal flows in gently, without splashing or turbulence. Once the metal appears in the riser (see Figure 3.7: here, the feeder doubles as the riser), filling is complete. After solidification, cooling, and de-molding, the entire "casting system" of sprue, runners, gates, and risers is separated from the actual part or product and recycled as production scrap. This recycling operation is common for all casting methods.

Because of the low product-specific investments, sand casting is well-suited for manufacturing of small series or even prototypes. However, despite the relatively long solidification times, the method can also work for high production volumes, simply because it is cheap to make and use multiple molds in parallel. In such an automated setup, a common example of which goes by the name of "Disamatic," cycle time is eventually only limited by the filling time.

## Gravity die casting

This method employs re-usable steel dies, which consist of two die halves separated by a parting plane, just as in sand casting. Compared to sand, steel conducts heat very well. The cycle times are therefore shorter, and the finer microstructure leads to better properties. Steel dies can also be finished smoothly, leading to equally smooth products with good control over all tolerances. Depending on the application, no further processing may be necessary except removing the casting system. As with sand casting, filling speeds are low—the metal is simply poured in under the force of gravity (hence the method's name)— and material properties can be very good. For this reason, the iconic "Moka Express" aluminum coffee maker is still made using this method, despite the millions that have been produced since its introduction in 1933. Like sand casting, gravity die casting allows disposable sand cores to realize hollow products. Product size and weight are more limited than for sand casting. Aluminum and magnesium are the most common metals to be cast with this method, brass much less so, and cast iron virtually never.

## Low-pressure die casting

Like gravity die casting, this method also employs re-usable steel dies, but it uses higher filling pressures, typically 2 to 5 bar. This is low enough to ensure laminar flows and therefore high material quality, but it allows excellent reproduction of die details (the pressure helps overcome the molten metal's surface tension, which otherwise would resist fine detailing). Apart from that, the method offers all of the benefits of the previous method. In industrial practice, the dies are often placed on top of the furnace containing the molten metal and filled through a riser tube using gas pressure. The metal column in the riser stays molten and acts as a feeder—a setup that is very different from gravity die casting. Again, aluminum and magnesium are the most used metals processed this way, with "alloy wheels" for cars being a common application.

## High-pressure die casting

With pressures up to 2,000 bar, this method certainly lives up to its name! Obviously, the steel dies that can withstand such forces are expensive: easily tens of thousands of euros for comparatively small castings (and typically 150% of the cost of a mold for plastic injection molding, for similar size and complexity: see also Chapter 8). Sand cores cannot be used and the method can therefore not produce products with nearly fully enclosed cavities like a jar or a flask, but it is possible to create local undercuts by using sliding elements in the die, as shown in Figure 3.8. Injection speeds are so high that material quality is poor due to turbulent flow and entrapped oxides. Still, this offers a huge advantage: it allows us to make thinner products, because the super-fast filling eliminates the problem of cold running. This means faster solidification and cooling, shorter cycle times, and higher productivity. Like the previous two methods, high-pressure die casting offers such good control over tolerances that we often do not need any further processing, apart from removing the casting system.

The range of metals is mainly limited to zinc, aluminum, and magnesium. When casting aluminum alloys, the die's life span is comparatively short. Nevertheless, alloys of this metal are commonly die cast, with Figure 3.9 providing a real-world example. Product size is another limitation, with 20 kg being a practical limit for part weight for most equipment. As for gate design, the method has an inherent trade-off: thin gates make it easy to separate the casting system from the product, but thicker gates

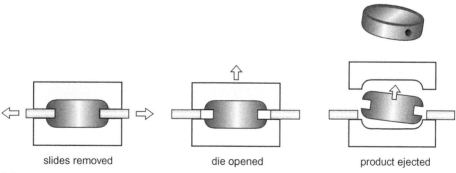

slides removed          die opened          product ejected

**FIGURE 3.8**

Creating an undercut with sliding mold or die elements. (For color version of this figure, the reader is referred to the online version of this chapter.)

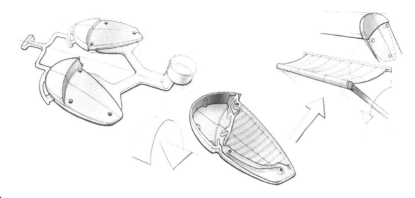

**FIGURE 3.9**

High pressure die-cast aluminum sunscreen cover plate (center) and casting system (left). Injection is done from the right side; the runners at the left side lead to a small reservoir for catching impurities.

that stay warm longer allow us to compensate for some shrinkage via the gates if sufficient "back pressure" is applied (i.e., keeping the pressure on after the mold has filled).

## Investment casting

In this special and still quite rare method, we use wax to produce a pattern. For high-volume production, several of these patterns are then attached to a vertical sprue (known in the trade as "the trunk"), much like the way leaves are attached to a tree branch. Next, this assembly is dipped in, or sprayed with, a slurry of ceramic particles and then sintered in an oven at high temperature. During this step, the slurry cures into a ceramic shell and the wax liquefies and flows out—for this reason, the method is also known literally as "lost wax casting," although the wax can usually be recycled. The resulting hollow ceramic shell is then used as the mold for the actual casting. Upon full solidification and cooling, the shell is destroyed through vibration and the castings are finished. The method is time-consuming and expensive but has a unique advantage: investment casting can produce extremely complex, high-quality parts with narrow tolerances ("near net shape"), with aircraft turbine blades being a prime example. Wax patterns can be made by joining as many individual parts as we want, and the restrictions inherent to parting planes and draft angles no longer apply. Solidification and cooling are slow, but this offers the advantage that we can accommodate relatively large variations in section thickness throughout one part. Furthermore, hardly any material is lost in the process, which makes this method ideal for expensive metals, such as the stainless cast irons. However, investment casting can also be used for aluminum parts that are too complex for other casting methods.

---

**EXERCISE 3.25**

Which method(s) would you choose for maximum design freedom? Think of the size of the casting and its material, but also of the ability to create undercuts, deliver fine detailing and sharp radii, or allow for changes in section thickness.

---

**EXERCISE 3.26**

Which method(s) would you choose for castings that are highly loaded and therefore require high strength and good ductility? And which one(s) would you choose in order to realize narrow tolerances?

---

**EXERCISE 3.27**

Which method wins when it comes to investments—that is, which is best for small production volumes? And which method offers the lowest part price, provided that production volumes are sufficiently large to amortize the mold or die costs?

---

## Guidelines for casting design

Designers enjoy having clear and meaningful "design guidelines": simple rules to ensure that the designs they come up with are easy to produce or, in this case, can be cast well, with high quality and against low cost—yet also with the required functionality. It is instructive to think of such rules for yourself! For instance, the more we keep the product's thickness constant, the less we have to worry about internal stresses or shrinkage porosity (see Section 3.5). And if we generously radius all corners, this improves metal flow while simultaneously avoiding stress concentrations. But take good notice: such rules vary in importance from method to method. For example, sharp variations in section thickness are much less of a problem in sand casting than in high-pressure die casting. Also, there are substantial differences between various metals: just compare the shrinkage of aluminum and cast iron. So ideally, our design guidelines must also aim at a specific metal.

---

**EXERCISE 3.28**

Look back at one of the product examples you located (see Exercise 3.1). Consider which metal and which method were involved, and then draw up three design guidelines for this combination. How are these rules reflected in your product?

---

## 3.8 **Further Reading**

The chapter has covered the manufacturing principle and main methods for metal shape casting, highlighting the most important attributes of function, quality, and costs. But, of course, we have not explored the field of shape casting fully. For instance, we only touched briefly on the topic of segregation and we have fully left out phase diagrams and the theories of phase transformation, which are ultimately indispensable for explaining what really happens during solidification. Designers who want to better understand casting of metals would do well to study further these concepts and their application to product design (e.g., alloy choice).

We have also limited ourselves to common casting methods and metals. Less common methods, such as *lost foam casting, centrifugal die casting*, the *Cosworth process,* and the *vacural process* (employed, for example, by Audi in the structure of their A8 luxury car) have been left unexplored. Likewise, several less common but important casting alloys, such as ductile iron, nickel superalloys, and precious metals, have not been mentioned. And there are also the processes of ingot, direct-chill, and continuous casting—the bulk processes used to make stock for wrought production by rolling, forging, and extrusion. If you would like to know more, refer to the following sources. The good news: you will find the concepts discussed in Sections 3.2 to 3.6 to be very useful to help you understand these other methods and metals. Such is the power of understanding the manufacturing principle.

For more academic and professional detail on casting, there are several journals, such as the bi-monthly *International Journal of Cast Metals Research.* For a more interactive exposure to casting technology and design, visit international casting trade fairs, such as GiFa in Europe (www.gifa.com) and the American Foundry Society in the United States (www.afsinc.org). For most readers, the following texts are likely to provide plenty of detail on different aspects of casting in design and engineering.

Ashby, M.F., Shercliff, H.R., Cebon, D., 2013. Materials: Engineering, Science, Processing and Design, third ed. Butterworth-Heinemann, Oxford, UK (Chapters 18 and 19). (*An introductory text at a complementary level to this book, providing detail on material properties and processing in a design-led context. It contains a "Guided Learning Unit" on phase diagrams and phase transformations, including solidification microstructures.*)

ASM Handbook, Volume 15: Casting, 2008. ASM International, Materials Park, Ohio, United States. (*The definitive, encyclopedic, reference on casting methods and materials, of biblical proportions. Other volumes deal with the other manufacturing principles in this book.*)

Campbell, J., 2003. Castings. Butterworth-Heinemann, Oxford, UK. (*Essential reading for anyone who wants to dig deeper.*)

Flemings, M.C., 1974. Solidification Processing (Materials Science & Engineering). McGraw-Hill, Ohio, United States. (*One of the truly seminal works. Out of print and expensive even as a second-hand copy, but definitely worth a good read.*)

# Sheet Metal Forming

4

## CHAPTER OUTLINE

This product is a support bracket for car starter units, to be mounted onto a garage wall. It is made of low carbon steel and is first rubber formed, then bent, and finally powder coated. Annual production volume: 500 units. Image courtesy Phoenix Metaal.

## 4.1 **INTRODUCTION**

Sheet metal forming lies right at the heart of industrial manufacturing. In almost all products that consist of multiple parts, at least one of these parts will be formed out of sheet metal, even if it is as simple as a bent strip of phosphor bronze to conduct electricity in a switch, or a piece of steel to act as a spring. Countless products are dominated by sheet metal parts—just think of cars—or are in fact almost nothing *but* such parts, as is the case for beverage cans. The main reasons for this popularity are that the sheet metal itself can be cheap, and forming and joining operations can be done rapidly. Annual production volumes can run into the hundreds of thousands for cars, or even hundreds of *millions* (e.g., cans, lamp fittings, or jar caps). Furthermore, sheet metal can be painted and textured to have many different kinds of look and feel, and it can provide a distinct product character, as seen in stainless steel kitchen utilities or classic car bodies. In short, it is an absolute *must* for any designer to know about this manufacturing principle.

Of course, sheet metal forming has its price, especially if we want to make complex, double-curved parts with close tolerances and accurate finishing. Such parts require high investment in methods called "matched die forming" and are only economically attractive if produced in large volumes. In contrast, the bending process, applicable to small volumes or prototypes, can be cheap but restricts our design freedom to single-curved parts. Between these extremes, many methods are available, each with its unique strengths and weaknesses.

To learn what these methods have in common we will consider the material science behind what happens when we deform a piece of sheet metal. This understanding will guide our design- or process-related choices, leading to the right synthesis of functional, quality, and economic demands. For reasons that will become apparent, we make a distinction between single-curved (Sections 4.2 to 4.5) and double-curved (Sections 4.6 to 4.10) parts and products, shown in Figure 4.1.

Note that sheet metal forming is always preceded by a process such as blanking or laser cutting to produce the right preform shape—such pre-cut pieces of sheet metal are referred to as "blanks." These processes can also be used as stand-alone operations, when only flat parts are needed. However, because they are based on different manufacturing principles, they are not discussed here. Interested readers are referred to the literature (see also Chapters 7 and 13).

---

**EXERCISE 4.1**

Look at home and find at least two examples of sheet metal parts or products. Distinguish between single- and double-curved parts and find at least one of each.

---

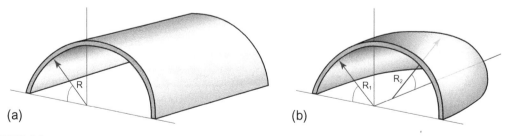

(a)                                                  (b)

**FIGURE 4.1**

Objects with (a) single-curvature and (b) double-curvature. (For color version of this figure, the reader is referred to the online version of this chapter.)

## 4.2 PLASTICITY, DISLOCATIONS, AND WORK HARDENING

Before we can discuss even a simple single-curved operation such as sheet metal bending, we must go through several basic concepts and get a feel for the magnitude of the properties involved. Figure 4.2 shows the well-known tensile stress-strain curve, drawn schematically for metals, with the four properties that are directly relevant to sheet metal forming: Young's modulus, yield stress, tensile strength, and strain-to-failure. Table 4.1 presents numerical values for these properties for various common

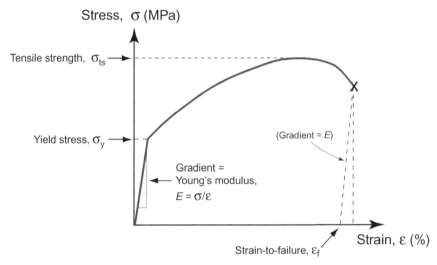

**FIGURE 4.2**

Tensile stress-strain curve for a typical metal (NB: elastic strain not drawn to scale). (For color version of this figure, the reader is referred to the online version of this chapter.)

**Table 4.1** Mechanical Properties of Various Sheet Metals

| Material | Young's Modulus [GPa] | Yield Stress [MPa] | Tensile Strength [MPa] | Strain-to-Failure [%] |
|---|---|---|---|---|
| Low carbon steel* | 200-215 | 250-395 | 345-580 | 26-47 |
| Low alloy steel* | 205-217 | 400-1500 | 550-1760 | 3-38 |
| Stainless steel (AISI 304) | 190-203 | 205-800 | 510-2240 | 5-60 |
| Aluminum alloy (5454-H12) | 71-74 | 195-215 | 250-280 | 13-15 |
| Aluminum alloy (6082-T6) | 70-74 | 240-290 | 280-340 | 5-11 |
| Copper (C12200) | 125-135 | 55-340 | 215-225 | 6-50 |
| Titanium alloy (Ti-6Al-4V) | 110-119 | 780-1080 | 860-1270 | 8-14 |

*Terms such as "low carbon steel" and "low alloy steel" refer to large groups of alloys and, consequently, a wide range of values for the respective properties. As a rule, it is only possible to give precise numbers for specific alloy types with specific rolling conditions or heat treatments.
Source: Cambridge Engineering Selector (CES), Granta Design, Cambridge, UK.

alloys. Whereas in structural design we are obliged to keep all stress levels below the yield stress, in sheet metal forming we need to exceed this stress level, because only in this way can we generate plastic strain and permanently deform the material. Note that sheet metals generally have some anisotropy, meaning that their properties (except the Young's modulus!) are different along the length, width, and thickness of the sheet metal, sometimes with differences up to 20% in value. However, to simplify our discussion, we will assume isotropic behavior.

---

**EXERCISE 4.2**

Which of these alloys has the "best" formability, and which has the "worst"? Explain briefly. Hint: consider which property is inextricably tied to tensile plastic deformation as a principle and which property can, at best, only be a limitation for the method or means of manufacturing.

---

**EXERCISE 4.3**

Choose a couple of alloys from Table 4.1, then use Hooke's law ($\sigma = E\varepsilon$) to calculate the maximum elastic strain in these alloys before plastic deformation begins. How does this value compare to the maximum plastic strain-to-failure?

---

In the elastic regime, the metal deforms by the interatomic bonds acting as linear elastic springs. On removal of the external force, all atoms spring back to their original positions, so we have no permanent deformation. Above a certain stress level—the yield stress—something radically different occurs: now there is sufficient force to activate and move *dislocations*. These are localized line imperfections in the crystal lattice. Under an imposed shear stress, the movement of dislocations along "slip planes" in the lattice allows incremental slip of one part of a lattice relative to another. Multiplied up millions of times over, on many different slip planes, this produces the macroscopic, and permanent, plastic deformation we observe. At room temperature (RT), deformation activates more and more dislocations. These hinder each other's movement and, consequently, the stress level required to maintain plastic deformation rises: the metal shows *strain hardening* (also known as *work hardening*). How strongly this happens differs in each metal.

The science of dislocations is a large, complex topic in materials science. For most design and engineering purposes though, we can focus on a small number of key ideas:

1. Because the dislocation motion involves blocks of material sliding past one another in shear, plastic deformation takes place at constant volume.
2. Beyond the yield point, the material may be able to strain permanently, but it does not suddenly become weaker—on the contrary, the strength rises further as it is deformed.
3. The elastic part of the strain remains throughout, and is still recovered on unloading (an important observation, controlling "springback" in sheet metal forming).
4. Dislocations impede one another, but so do many other microstructural features within the lattice (e.g., alloying elements in the form of solid solution, second phase particles). This explains the huge diversity in strength, even within one class of alloys—but the key properties of Figure 4.2 capture the differences in a small number of parameters.

To get a deeper understanding of plastic deformation, dislocations, and alloying, see the standard texts listed under Further Reading.

Strain hardening does not have to be permanent. Heating the metal to a sufficiently high temperature can significantly reduce the number of dislocations, by the mechanisms of *recovery* and *recrystallization*. This "annealing" restores the original (lower) yield stress and (higher) strain-to-failure. It takes extra time and effort, but it is sometimes necessary if we want to make complex sheet metal products. Note that the deformations cannot be undone this way!

---

**EXERCISE 4.4**

Consider Figure 4.2, and suppose we deform the metal to a stress halfway between its yield stress and its tensile strength and then unload it. On a sketch of the stress-strain curve, show how you would find the permanent strain up to this point. The sample is then re-loaded to find its new, strain-hardened yield stress. Explain why the value will be higher than the nominal value at the point when unloading was started in the previous test. Hint: think carefully about how the sample dimensions change during a tensile test, and how yield stress is defined.

---

**EXERCISE 4.5**

At first sight, Figure 4.2 shows something peculiar: the material apparently fails at a stress level below its tensile strength. Explain this mystery. Hint: consider how the stress level is defined and measured during tensile testing.

---

## 4.3 THE MINIMUM BENDING RADIUS

In the most basic of all sheet metal forming operations, a flat blank of constant thickness $t$ is bent over an angle $\alpha$ with a radius $R$ (Figure 4.3). Exactly how this can be done will be discussed in Section 4.5, but what concerns us now is: how small can $R$ be, as a function of $t$, $\alpha$, and the material's formability? To answer this question, we draw up a simple model by making these five assumptions:

1. All deformation that takes place in a zone defined by the angle $\alpha$ and the radius $R$ is assumed to be constant. Outside this zone, no deformation occurs.
2. The metal elongates in tension in the same manner as in compression (i.e., the stress-strain curve is anti-symmetric with respect to the origin).
3. At mid-thickness of the sheet, at the radius equal to $R$, there is a "neutral axis" (with length $\alpha R$; $\alpha$ in radians) that does not change in length.
4. The metal outside the neutral axis elongates and becomes thinner, and the metal inside the neutral axis compresses and becomes thicker.
5. The metal is isotropic and homogenous, meaning that is has the same behavior in all directions and at all locations.

The result of these assumptions is a bending model in which the distribution of strain over the cross-section depends linearly on the distance from the neutral axis. The maximum strains occur at a radius of

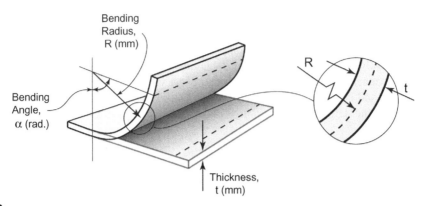

**FIGURE 4.3**

Key parameters in the bending process. (Inset: detail of cross-section.) (For color version of this figure, the reader is referred to the online version of this chapter.)

$R \pm t/2$ (i.e., at the sheet surfaces). In principle we want maximum design freedom, so is there a limit to the radius and angle that can be formed from a given thickness? The tensile strain-to-failure $\varepsilon_{max}$ sets such a limit, captured in this context in terms of a *minimum bending radius* (MBR).

---

**EXERCISE 4.6**

Derive a generic formula for the MBR as a function of geometry and formability, assuming that the strain-to-failure in bending is the same as it is in a tensile test. (Hint: how much longer does the outermost fiber of the bending zone get, relative to the length of the neutral axis?) Choose a couple of alloys from Table 4.1, and calculate the MBR for a bending angle $\alpha = 60$ degrees ($= \frac{1}{3}\pi$) and a sheet thickness $t = 3$ mm.

---

**EXERCISE 4.7**

In his steel bookshelf *Storyline* (Figure 4.4), Dutch designer Frederik Roijé has made elegant use of the bending process. What would this product look like if he had chosen aluminum instead? Bonus question: Would it then be lighter, and if so, how much?

---

This simple model allows us to determine the MBR not just for sheet metal but also for solid bars, rods, and beams. For instance, we can quickly determine whether the design for a chair with sharply bent circular metal legs is feasible (in the formula, the tube diameter $D$ then replaces the sheet thickness $t$). Care is needed for bending of hollow tubes though—these are likely to fail first by buckling (think of bending a drinking straw). Incidentally, we should note that the real Storyline shelf has local incisions at the corners. This allowed for an even sharper and book-friendly design.

---

**EXERCISE 4.8**

Look closely at the single-curved product(s) you found in Exercise 4.1. What is the sheet thickness, and what is (approximately) the bending radius? What does this tell you about the material that was used?

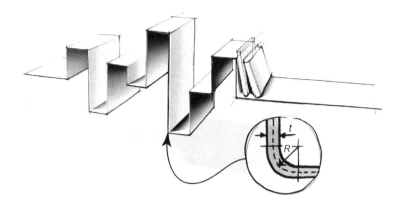

**FIGURE 4.4**

"*Storyline*" bookshelf. (Inset: radius detail.) (Design: Frederik Roijé)

In practice, we can often bend metals quite a bit sharper than this simple model predicts. There are three main reasons why the model is too conservative. First, consider the edges of the bending zone, and note that it is unrealistic to expect that the metal just inside this zone deforms, while the immediately adjacent metal just outside this zone does not. Strain hardening simply does not allow such a sharp boundary between deformed and non-deformed metal. So there will be a transition zone between the fully deformed radius and adjoining undeformed material—the length of these zones is of the order of the sheet thickness. The result is that we can bend sharper than predicted by the first model.

---

**EXERCISE 4.9**

Make a clear sketch of the transition zones. Do you think they get more influential in reducing the MBR for small bending angles than for large ones, or vice versa? Explain.

---

The second reason is that metals behave differently during tensile testing than during bending. Under uniaxial tension, a metal test bar shows "necking" after the stress reaches its maximum value (the tensile strength)—that is, its width and thickness reduce significantly at a localized cross-section, prior to tensile failure. This occurs because the material is no longer able to harden fast enough to maintain uniform deformation. In bending, the same degree of strain hardening will have occurred on the outer surface (the tensile side), but necking is inhibited by the material immediately below, in which the strain is smaller. The result is again that we can bend a bit sharper than tensile test data suggests.

For the third reason, we return once more to the stress-strain curve. After necking occurs, the deformation is concentrated in a small volume, as just mentioned. But in calculating the strain-to-failure, we still divide by the original sample length $L_0$ (definition: $\varepsilon_{max} = (L - L_0)/L_0 = \Delta L/L_0$), which is much longer than the length that is actually being deformed, so *locally* the strain is much higher. The strain to failure underestimates what the material really can do in terms of bending formability. Hence, we can usually bend sharper than we would expect on the basis of $\varepsilon_{max}$ alone. This effect is especially relevant for materials that exhibit "diffuse necking." Chief among these is low carbon steel; aluminum, for instance, is much less affected by it. A word of caution, however: strain-to-failure is *not* a strict material property, as the necking contribution to $\varepsilon_{max}$ increases with the sample thickness.

---

**EXERCISE 4.10**

Return to Exercise 4.2. How would you now refine your original answer?

---

## 4.4 SPRINGBACK

Recall from the stress-strain curve that the elastic strain associated with the applied load remains throughout any plastic deformation, until the part is unloaded. In sheet metal forming, we therefore always have to deal with elastic *springback* (i.e., a change in shape of the part as the load is removed). Furthermore, springback leads to internal *residual stresses* over the cross-section, distributed such that they balance each other out, as equilibrium requires that there is no net force (Figure 4.5).

If we could predict exactly how much springback there will be, then we could account for it by deforming a bit further. For instance, if for a 90-degree angle we knew beforehand that there will be 5-degree springback, then we could bend to 95 degrees and end up with what we want (we could, of course, also use *trial-and-error*, but that would require time and effort that we would rather avoid). In our predictions, we can make good use of the concept of "deformation energy." Sheet metal forming requires work: we have to apply a certain force over a certain distance, and force ($N$) times distance ($m$) equals work ($N\,m = J$), which is a form of energy. Now force per square meter equals stress ($N/m^2$) and distance per meter equals strain ($m/m$, or %). From this information we find that stress times strain equals energy too, but then per unit of volume ($J/m^3$). This deformation energy is equal to the area under the stress-strain curve up to the point reached during the deformation process.

---

**EXERCISE 4.11**

Select any alloy from Table 4.1 and estimate the maximum elastic energy that this metal can store, as well as the maximum plastic energy that can be absorbed, per unit volume. To facilitate the calculation, assume a stress level in the plastic region that is equal to the mean of the yield stress and the tensile strength.

---

As it happens, the ratio between elastic and total (i.e., elastic plus plastic) deformation energy is a good indicator for the amount of springback during bending. If this ratio doubles, we will get roughly twice the springback. Hence, to minimize springback, we must minimize the energy ratio, which effectively means we must maximize plastic deformation. This, of course, does not yet tell us the absolute amount of springback (that, sadly, is difficult to predict), but it does get us under way: once we know the springback in a suitable reference case, we can compare the deformation energies to estimate what to expect in a related case. For instance, we can change the thickness $t$ while keeping bending radius $R$ and material (i.e., $\varepsilon_{max}$) the same, or some other combination (note that we should also assume that the bending method is the same as well—see Section 4.5).

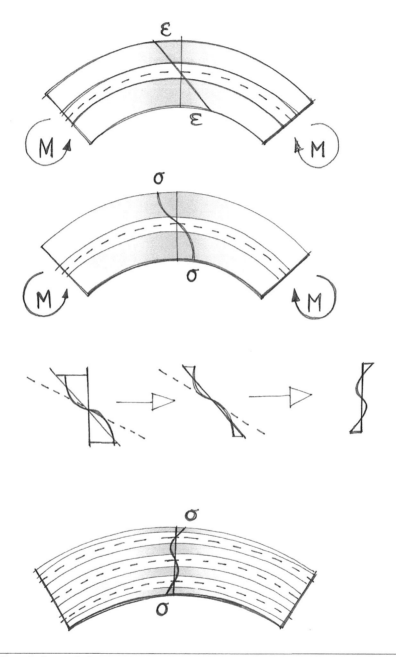

**FIGURE 4.5**

Strain distributions during and after bending, in cross-section.

---

**EXERCISE 4.12**

Assume that during bending of a certain steel computer casing we get 2-degree springback. What will the springback be if we double the bending radius? And what if we exchange the steel for aluminum (keeping the thickness the same)?

---

**EXERCISE 4.13**

Do you think that there was a lot of springback during the manufacture of your single-curved product(s) from Exercise 4.1, or not? How can you tell?

---

**EXERCISE 4.14**

For car body panels, dimensional tolerances are tight, and springback must be tightly controlled. Use this information to explain why the car industry allows only small variations in the yield stress of any sheet metal it buys. And could the same tooling be used if the manufacturer changed from steel to aluminum of the same strength and thickness, in exactly the same shape?

---

In passing, it is worth mentioning that automotive crash protection structures also operate by absorbing energy through plastic deformation. Moving vehicles crashing to a stop must dissipate all of their kinetic energy safely. So for the structure to absorb this energy, we need an alloy with high energy absorption potential per unit volume *and* a large enough volume of this material being deformed during the crash. This is the basic idea behind "crumple zones" in cars.

---

**EXERCISE 4.15**

A certain high strength steel (HSS) can absorb 200 MJ/m$^3$ of deformation energy. How many kg of such HSS must be fully deformed to bring a 1500 kg car moving at 100 km/h to a standstill? Hint: assume the density of the steel is 7,900 kg/m$^3$.

---

## 4.5 MANUFACTURING METHODS FOR SINGLE-CURVED PARTS

We round off the topic of single-curved sheet metal forming with a discussion of three manufacturing methods: bending, roll bending, and roll forming. The first two can, in suitable variations, also be used to process solid bars, hollow tubes, or profiles. So the concepts of MBR and springback can also be applied, for example, to the chrome-plated bent steel tubes of the famous Dutch Gispen office furniture, and to countless other products.

## Bending

From simple, universal equipment suitable for prototyping to fully automated installations dedicated to high-volume manufacture, this method can be found anywhere we need to make one or more straight bend lines in a sheet metal part. In principle, all metals—indeed, all materials showing yield behavior—can be bent. Provided that the sheet thickness and the material's yield stress are well-controlled, we can control springback very well and deliver accurate results, if not by prediction then at least through trial-and-error. Obviously, form freedom is set by the upper limits on part size, thickness, and material strength, which are governed by the equipment that is industrially available. As for costs, investments, cycle times, and the like, it is difficult to provide specific guidance—there is a huge variety in bending equipment, and bending is often combined with many other operations, such as cutting or punching. But in comparison to other forming methods, bending is relatively cheap and fast.

Bending can be done in several ways. For instance, in *air bending* (Figure 4.6a), we deform the blank by means of three-point loading, supporting the ends and deforming at a small central contact area. Conversely, in *bottoming* or *V-die bending* (Figure 4.6b), we completely enclose the blank between two profiled metal dies, ensuring that deformation is more homogeneous and with less springback than in air bending. A third variety is *wipe bending* (Figure 4.6c), in which the blank is first gripped between

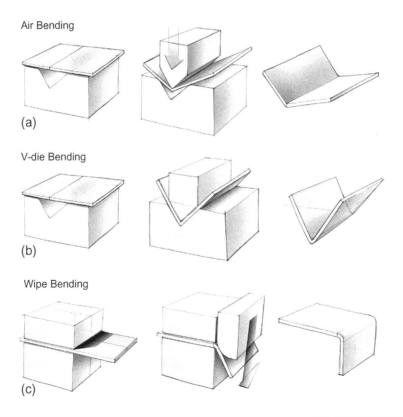

Air Bending

(a)

V-die Bending

(b)

Wipe Bending

(c)

**FIGURE 4.6**

(a) Air bending, (b) V-die bending, and (c) wipe bending. In each case, the processes are shown before, during, and after.

two clamps and then bent as a cantilever by a moving die. *Stretch bending* (not illustrated) is a fourth, in which the edges are clamped as the transverse load is applied, and there are more. The key to understanding how they work is analyzing the forces acting on the blank, as the material will always begin to deform where the bending moment is greatest.

## Roll bending

If bending is well-suited to making relatively sharp radii, then roll bending is ideal if we want to make larger ones (Figure 4.7). This method allows us to make a cylindrical product, or even, with the right kind of control over the rollers, cones. The forces required to do this are relatively modest, especially because the deformation is usually obtained gradually: the blank passes back and forth through the machine several times, with the rolls being repositioned after each pass. Because the radii are large, springback presents a challenge, as the ratio of elastic over total deformation energy is relatively high.

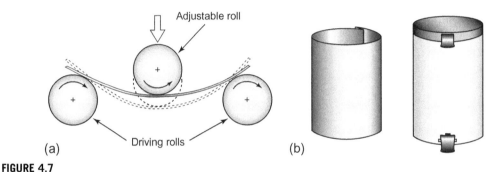

**FIGURE 4.7**

(a) Schematic of roll bending; (b) curved strip and rolled product (pedal bin). (For color version of this figure, the reader is referred to the online version of this chapter.)

## Roll forming

In this relatively unknown but valuable process, a long strip of metal is fed through a series of rollers at considerable speed—sometimes more than 1 m/s—to produce a profile (Figure 4.8). The rollers progressively generate the desired cross-section, which can be surprisingly varied in shape. From thin office furniture frames to heavy-duty truck chassis beams, roll-formed products are more common than you might think. The method is robust and gives tight control over tolerances, but the product-specific investments are high and it is therefore only economically attractive for high volumes (not to mention the time needed to set up the entire process). Maximum strip width—typically 0.5 m, depending on the metal in question—puts another restriction on the design freedom. However, many common secondary operations, such as cutting to length or hole-punching, can be put in line with the roll forming process. By integrating a welding operation at the end, it is possible to manufacture closed cross-sections. Furthermore, industrial roll formers offer a wide range of standard profiles, giving effective solutions to many common design problems. Roll forming is mostly done with low to medium carbon steels, certain low alloy steels, and stainless steels (notably the common AISI 304), but typically not with aluminum (which is more commonly extruded to make shaped profiles: see Chapter 5).

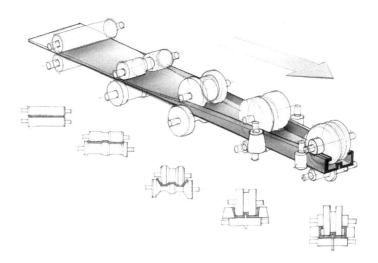

**FIGURE 4.8**

Schematic of roll forming, with resulting profile.

---

**EXERCISE 4.16**

Place the three manufacturing methods discussed previously in order of form freedom (meaning part size, sheet thickness, material choice, and shape complexity). Next, place them in order of flexibility: how easy is it to make small changes to parts? What do you conclude?

**EXERCISE 4.17**

Think of a sensible application that combines two of these three methods (e.g., roll forming and roll bending).

**EXERCISE 4.18**

Is it possible—without raising the temperature!—to bend glass, or roll form a strip of plastic, such as polyethylene (PE)? And what about rolling plywood? Explain your answer.

Design guidelines for single-curved sheet metal forming are easy enough to find, but they only become truly useful when they are detailed. For instance, one could state: "To minimize springback, make the bending radius as small as possible," but this is only effective if it includes the formula for the MBR, as derived in Exercise 4.6. Then for a specific material and thickness, we can find a numerical value for our design.

---

**EXERCISE 4.19**

Propose two additional design guidelines for single-curved metal parts. Ensure that they are specific and/or that they refer explicitly to the theory covered in Sections 4.2 to 4.4. Look back at your sample product(s) from Exercise 4.1: are your guidelines reflected in these parts?

---

## 4.6 DOUBLE-CURVED PARTS: THE CONCEPT OF TRUE STRAIN

Single-curved parts and products may be comparatively easy to understand and realize, but for many applications, their form freedom is too limited—many designs require double curvature. To understand methods to form double-curved products, we must expand our knowledge of stresses and strains, including the concept of true strain.

For convenience, strain in material tensile testing is usually defined as *nominal strain, $\varepsilon_n$*, the change in length $(L - L_0)$ divided by the original length, $L_0$ (often expressed as a percentage):

$$\varepsilon_n = (L - L_0)/L_0 = \Delta L/L_0 \tag{4.1}$$

All strains-to-failure listed in Table 4.1 are nominal strains. In metal forming, however, it is often more convenient to use *true strain, $\varepsilon_t$*, in which each increment of extension is related to the instantaneous length, integrated over the full deformation:

$$\varepsilon_t = \ln (L/L_0) \tag{4.2}$$

True strains can be substantial and are usually expressed as a (dimensionless) number, rather than a percentage, which serves as a handy distinction between the two. Combining these two equations, we find the relationship between nominal and true strain (taking care to use absolute numbers for both, rather than percentages):

$$\varepsilon_t = \ln (\varepsilon_n + 1) \tag{4.3}$$

The logarithmic form of true strain has certain advantages, as we will see.

---

**EXERCISE 4.20**

If we have a nominal strain of 50%, then what is the accompanying true strain? And how much is a true strain of −0.16, expressed as a nominal strain?

---

**EXERCISE 4.21**

Suppose we stretch a piece of sheet metal first 30%, then another 20%, and finally another 10%, all in the same direction. What is the total nominal strain after these three steps?

Again we deform in three steps, now with true strains of 0.30, 0.20, and 0.10. What is the total true strain after three steps? And what is the difference—or more to the point, the advantage—compared to the calculation you made in Exercise 4.21?

When we deform metals in multiple steps, it is easier to find the final true strain than to find the final nominal strain: we can simply add up the strains at each step. For some alloys we can look up the allowable maximum true strains in engineering handbooks, although these data are considerably scarcer than nominal values.

Forming double curvatures is an inherently multi-axial process, so we need to expand our understanding of multi-axial stress and strain. Figure 4.9 depicts a small piece of sheet metal. As we know, a tensile stress $\sigma_1$ applied in one direction generates an elastic strain in that direction, $\varepsilon_1$. At the same time, we also get elastic contraction in the other two directions, giving (negative) strains in these directions, $\varepsilon_2$ and $\varepsilon_3$. In an isotropic material, these lateral strains are equal and scale with the applied strain:

$$\varepsilon_2 = \varepsilon_3 = -\nu \varepsilon_1 \tag{4.4}$$

where the Greek letter $\nu$ (pronounced as "nu") is *Poisson's ratio*. Note therefore that even though the stress is uniaxial stress, we have strains in all three directions!

For elastic deformation, $\nu$ lies in the range 0.25 to 0.35, with the exact value depending on the alloy. This implies that during elastic deformation, the volume of the material changes a little bit. For plastic deformation, however, the volume remains constant (due to the mechanism of dislocation motion, discussed earlier). This turns out to correspond to a value of Poisson's ratio $\nu = 0.5$. So for plastic deformation, we only need to know two strain components to be able to derive the third, because for $\Delta V = 0$, we find that $\varepsilon_1 + \varepsilon_2 + \varepsilon_3 = 0$ (we leave the proof as an exercise). Note that this is only valid for true strains, not for nominal strains (unless they are very small), which is another advantage of working with true strain.

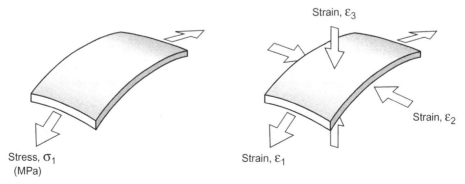

**FIGURE 4.9**

Stress (left) and accompanying principal strains (right) on a piece of sheet metal.

---

**EXERCISE 4.23**

Show that during plastic deformation, when the volume remains constant ($\Delta V=0$), the sum of the true strains must be zero: $\varepsilon_1+\varepsilon_2+\varepsilon_3=0$. Hint: consider a rectangular block of initial dimensions ($L_0, B_0, H_0$) and final dimensions ($L_1, B_1, H_1$), noting that $L_0\,B_0\,H_0=L_1\,B_1\,H_1$. Start by writing down the true strains in terms of the dimensions.

---

**EXERCISE 4.24**

Explain how the behavior of dislocations in metals means that plastic deformation takes place at constant volume.

---

## 4.7 THE FORMING LIMIT DIAGRAM (FLD)

To make double-curved sheet metal parts, we must exert a force in at least two different directions. This can be tension in combination with either tension or compression (with the magnitude of the latter being of lower or equal magnitude to the principal tension). In these situations, the applied stress is no longer uniaxial, but *biaxial*. Consequently, data from uniaxial tensile tests are no longer sufficient to determine the formability limits. Instead, we use a clever concept: the *forming limit diagram*, or FLD (Figure 4.10).

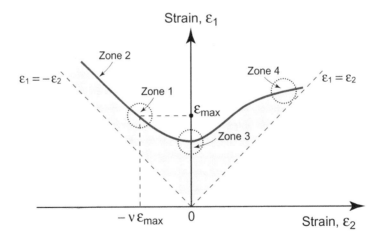

**FIGURE 4.10**

Forming limit diagram, plotting allowable strain $\varepsilon_1$ (vertical) against $\varepsilon_2$ (horizontal). (For color version of this figure, the reader is referred to the online version of this chapter.)

This diagram shows the viable combinations of biaxial strain $\varepsilon_1$ and $\varepsilon_2$. These are the in-plane strain components in the sheet in the tensile direction ($\varepsilon_1$) and along the perpendicular direction ($\varepsilon_2$), which may be tensile or compressive. Because $\varepsilon_1$ must be greater or equal to the absolute magnitude of $\varepsilon_2$, the area below the diagonals $\varepsilon_1 = \varepsilon_2$ and $\varepsilon_1 = -\varepsilon_2$ does not apply (with the convention that $\varepsilon_1$ is the principal tensile direction). The diagram is limited at the top by the *forming limit curve* (FLC), above which the material will fail. In principle, we can go from the origin to any point in the gray area by some kind of forming process. The strains are given as true strains, and therefore we can at each point also find the accompanying strain $\varepsilon_3$ in the third direction—through the thickness—via the equation $\varepsilon_1 + \varepsilon_2 + \varepsilon_3 = 0$.

First consider the strains imposed under uniaxial stress. Recall that the strain is multi-axial, with a strain-to-failure $\varepsilon_1 = \varepsilon_{max}$. The transverse strain under homogeneous deformation will be $\varepsilon_2 = 0.5\,\varepsilon_{max}$ (i.e., taking Poisson's ratio $v = 0.5$ and neglecting the necking strain). This is indicated as zone 1 on the FLC. By imposing a greater compressive strain, the limiting tensile strain-to-failure can be increased above $\varepsilon_{max}$ (zone 2). But if the transverse strain is prevented and is around zero, the maximum value for $\varepsilon_1$ becomes substantially lower than $\varepsilon_{max}$ (zone 3). What is perhaps most surprising is that, on the right of the diagram, the combination of tensile strains in both directions also leads to a limiting tensile strain above $\varepsilon_{max}$ (zone 4). As a rule-of-thumb, we may typically assume that the maximum value for $\varepsilon_1$ is between 1.6 and $2 \times \varepsilon_{max}$.

---

**EXERCISE 4.25**

Refer to the FLC in Figure 4.10 and estimate the through-thickness strain ($\varepsilon_3$) along the curve. Where is this strain component a maximum?

---

**EXERCISE 4.26**

Make a sketch of a small, rectangular plate element. Next, sketch what this element will look like after undergoing the combined strains of tension-tension (zone 4), tension-near-zero strain (zone 3), tension-compression (zone 2), and uniaxial stress (zone 1).

---

**EXERCISE 4.27**

Revisit Exercise 4.2 once more. What is your answer now?

---

As a final note on the FLD, we return to single curved bending. Along the bend line, the width of the sheet resists lateral contraction, forcing $\varepsilon_2$ to be close to zero. Consequently, the FLD shows that the strain-to-failure $\varepsilon_{max}$ overestimates the MBR we can achieve, as bending takes place not in zone 1 but in zone 3. Clearly, single curved bending is not as simple as it looks.

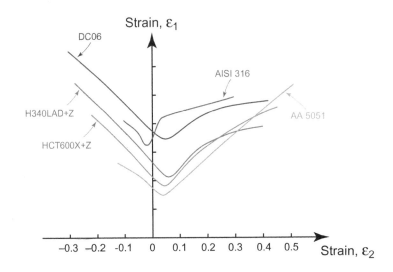

**FIGURE 4.11**

Actual FLCs for various metals:

- DC06: an IF ("interstitial-free"), high ductility, low strength deep drawing steel
- H340LAD+Z: a cold rolled, galvanized automotive steel with 340 MPa yield stress
- HCT600X+Z: a cold rolled, galvanized automotive steel with 600 MPa tensile strength
- AISI 316: an austenitic stainless steel suitable for welding
- AA 5051: a medium-strength aluminum alloy

(For color version of this figure, the reader is referred to the online version of this chapter.)

Figure 4.11 shows typical forming limit diagrams for a range of sheet metals. The ranking of ductility of the alloys is as expected along the uniaxial $\varepsilon_1$ axis, but the biaxial formability behavior away from this vertical axis shifts the relative performance.

## 4.8 APPLICATION OF THE FLD: DEEP DRAWING

One of the simplest of all double-curved products is a cylindrical beaker. The concept of the FLD can be instructively applied to how such a product can be manufactured. Figure 4.12 shows, in three steps, how a beaker can be formed out of a cylindrical blank. The beaker has diameter $D_b$ and depth $H_b$; the blank initially has diameter $D_p$. The double-curved zones are found at the transitions from flange to sidewall and from sidewall to bottom. The detail of how we produce such a shape will be discussed later in this section, but for now we simply assume that we press a punch die through the blank, with the outer shape of the die matching the inner shape of the beaker.

Consider the blank to be divided into four areas labeled A, B, C, and D, with representative points marked along a radial line (see Figure 4.12). At the start of the forming process, the circular area A lies in the middle of the blank, whereas areas B, C, and D lie as concentric rings out toward the edge. During forming, these four areas experience different combinations of strains:

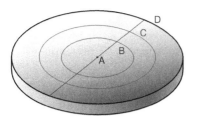
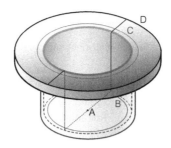
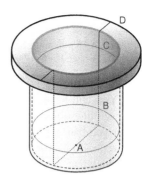

**FIGURE 4.12**

Three stages of deep drawing of a circular beaker. (For color version of this figure, the reader is referred to the online version of this chapter.)

**A.** Throughout the process, this area is elongated in the radial direction $\varepsilon_1$ as well as in the tangential (circumferential) direction $\varepsilon_2$—a combination of biaxial tensile strains known as "stretching."

**B.** This area first sees elongation $\varepsilon_1$ and $\varepsilon_2$ (in the radial and circumferential directions), as A; but later, it gets pulled "over the edge" of the base and changes to a combination of elongation $\varepsilon_1$ (along the length of the beaker) plus zero strain $\varepsilon_2$ (around the circumference).

**C.** This area first sees elongation $\varepsilon_1$ (in the radial direction) plus compression $\varepsilon_2$ (in the circumferential direction), referred to as the "deep drawing strains." Later, it gets pulled "over the edge" at the top, and changes to a combination of elongation $\varepsilon_1$ (along the depth of the beaker) plus zero strain $\varepsilon_2$ (around the circumference); the farther away that the ring lies from the edge of the beaker, the later the transition in strain path.

**D.** Throughout the process, this area undergoes deep drawing strains.

---

**EXERCISE 4.28**

In an FLD, sketch the trajectories that the four areas A to D travel through, each one beginning at the origin. Which area will reach the FLC (wherever that curve may be) the soonest, and what does that mean for the manufacture of this beaker?

---

**EXERCISE 4.29**

Determine how much a small segment of area D must be compressed if it is to end up at the edge of the beaker, starting at the edge of the blank. Hint: its arc length before and after deformation is proportional to its "old" and "new" diameter, and you may assume that the thickness remains constant. If the tangential radial compression is $\varepsilon_2$, then how large is the radial strain $\varepsilon_1$?

---

**EXERCISE 4.30**

Assuming now that the bottom does not deform at all, determine all three strain components $\varepsilon_1$, $\varepsilon_2$, and $\varepsilon_3$ in the wall of the beaker. Hint: $\Delta V = 0$.

---

You will have noticed that you need to make several clever assumptions and simplifications to answer questions like the previous two. For starters, if the product is rotationally symmetrical, then you can assume that all strains must show the same symmetry (provided that the material is isotropic—a useful assumption, but one that unfortunately is not quite true in normal practice). Also, consider that because the beaker's circumference is constant, the strains along its circumference must be equal to zero. However, this is not the case in the flange, where these tangential strains are negative. Finally, consider that the volume cannot change; for the wall of the beaker this means that the depth $H_b$ depends on the strain across the thickness $\varepsilon_3$.

---

**EXERCISE 4.31**

Choose realistic values for the dimensions of the beaker and blank, then select one alloy from Figure 4.11. Can this particular beaker be produced from this metal? Explain.

---

Note that the material in area C experiences another deformation when it slides from the flange over the edge of the beaker into the sidewall: when this happens, it gets bent out-of-plane as well as deformed within its plane. Seen in cross-section, we get a combination of bending with stretching, with more strain in the radial direction on the outside of the radius, and less strain on the inside. Of course, this also happens in area B at the transition from bottom to sidewall.

---

**EXERCISE 4.32**

Calculate the extra strains due to this out-of-plane bending, if the bending radius is given as five times the blank thickness. Give your answer as a true strain. Exactly where (over the thickness of the material) do these extra strains occur? Are they significant?

---

Let us now consider briefly how such a beaker can be made. In practice, blanks can be compressed only very little before they begin to buckle. For this reason, we need a "blank holder"—that is, a heavy steel ring placed concentrically around the punch die and pressing down onto the blank (see also Section 4.10). A lubricant ensures that the blank can slide inward from under the blank holder and deform. (Incidentally, this also keeps the material from getting thicker, which is why for Exercise 4.30 we could assume that $\varepsilon_3$ in the flange is zero.) Still, all deformation in the flange is driven by the drawing force in the beaker wall, generated by the punch die pushing on the bottom. If the flange is too large, the wall of the beaker will fail, and for this reason, very deep (or thin) products are produced in multiple forming steps, with a separate die for each step, getting successively smaller in diameter (in other words, this is done working from the outside inward).

---

**EXERCISE 4.33**

Take a closer look at your double-curved product sample(s) from Exercise 4.1. Which areas will have experienced deep drawing, and which ones stretching? Which areas have seen much deformation, and which only little? Any new thoughts on the material that has been used?

---

A final note on a practical difficulty with deep drawing, particularly for thin-walled products such as beverage cans. We have assumed isotropic behavior—the same yield response in all directions. But blanks for circular products are stamped out of rolled sheet, and because of the rolling process the sheets show *anisotropy*—different strengths along and across the sheet. A common source of problems in deep drawing rolled sheet is that the outer edge of the blank goes wavy during deep drawing—a phenomenon known as "earing". The underlying reason is that the slip planes in the grains are not randomly arranged, but rolling has forced the crystals to become oriented in more specific patterns with respect to the axes of the sheet. The statistical spread of orientations can be measured and mapped—it is called *crystallographic texture*. Surprisingly, annealing heat treatments to fully recrystallize rolled sheet do not eliminate texture, but a different anisotropy emerges. Texture, and the resulting anisotropy, depend in complex ways on the alloy composition, on the rolling and annealing conditions, and on all sorts of microstructural detail inherited from the original casting and other previous processes. It may not look it, but in metallurgical terms the thin-walled beverage can is one of the most heavily researched and sophisticated products.

## 4.9 SPRINGBACK IN DOUBLE-CURVED PRODUCTS

Much like single-curved products, double-curved products can exhibit springback. This effect now has two components: springback out of plane and springback within the plane. The first component can, in principle, be determined using the concept of deformation energy introduced in Section 4.4. The second component is the in-plane elastic strain, which we know from Exercise 4.3 to be very small. As before, we can state that springback will always counteract part of the deformation imposed on the blank, but beyond that, it is difficult to make any kind of quantitative prediction.

In products that have strong double curvature, the product shape itself is so stiff that springback is largely prevented. If the product happens to be symmetrical, then we usually cannot see any springback at all. This does not mean the problem does not exist: instead of springback, we will end up with internal stresses in the product (this can be compared to the thermally induced stresses in metal castings, which cause either distortion or internal stresses). Imagine we deform a round blank to create a hemispherical bowl (Figure 4.13b). The bowl wants to get flatter and wider, but this is resisted by the stiffness of the shape.

---

**EXERCISE 4.34**

What kind of springback do you expect in producing a shallower product, such as the Alessi bowl depicted in Figure 4.13a, and why?

---

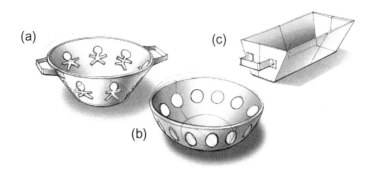

**FIGURE 4.13**

(a) Alessi bowl; (b) hemispherical bowl; and (c) square baking tray.

---

**EXERCISE 4.35**

In terms of the deformations that are required to make it, the rectangular tray shown in Figure 4.13c is not a difficult product. However, what about springback? More specifically, do you expect the sides to remain straight, or to curve inwards or outwards due to springback?

---

**EXERCISE 4.36**

What do you recommend for making the Alessi bowl: first forming, then stamping out the figurines, or vice versa? Explain.

---

In shallow products (where the elastic deformation energy we put in is relatively large compared to the plastic deformation energy), having some kind of flange or other kind of stiff reinforcement around the edge is crucial for controlling springback. And to make the most of the situation, the flange can then be used also for assembly purposes. In automotive *bodies-in-white* it is easy to see and study such flanges.

---

## 4.10 MANUFACTURING METHODS FOR FORMING DOUBLE-CURVED PARTS

We conclude our exposition of double-curved sheet metal forming with some remarks on six interesting manufacturing methods: panel beating, deep drawing, matched die forming, sequential die forming, rubber forming, and hydroforming. The methods are distinguished here mainly by considering the number of part-specific investments (a secondary aspect could be to consider how the force required for forming is exerted).

In all of these processes, a certain amount of material is turned into scrap, because not all of the ingoing material can end up in the part (blanking, edge trimming, etc.). For certain parts, the scrap percentage can easily be 50%. However, in large-scale industrial practice, this scrap can always be recycled, and usually with little or no loss in quality.

## Panel beating (hammering)

For manufacture of prototypes or small series of double-curved parts, panel beating is a logical choice. This method operates by manually placing a blank between two stamps, each shaped like a hammerhead. The two dies rapidly move up and down forcefully, "beating" the blank between them. In the process, the yield stress in the direction of the blank's thickness is exceeded in compression, creating positive strains within the blank's plane ($\Delta V = 0$). By beating the blank in different places in succession, it is possible to create complex parts. This requires great skill and experience, and of course it is time-consuming, but apart from a wooden die with which one can check if the part has the required shape, there are no part-specific investments. Parts made by panel beating are characterized by their flowing, organic shapes with generous radii. Many classical car bodies, such as the well-known Porsche 911, were originally made by this method. However, it is still very much alive, and some researchers have even worked out how to automate the method, complete with software to make predictions of formability.

## Deep drawing

This method is suited for series production of all kinds of products, ranging from beakers and cans to cooking pots. The product-specific investments consist of a drawing die (or simply "die"), a blank holder, and a punch die (or simply "punch"; Figure 4.14). Initially, the blank is clamped between the drawing die and the blank holder. Then, the punch is pressed down, forming the product. The drawing die determines the maximum outer diameter, the punch the minimum inner diameter. Note that beneath the punch, the blank is free—a key difference with the next method. Deeper products can be made in successive forming steps, each requiring their own tools.

Observe closely how the deformation takes place. Initially, the punch creates an indentation in the blank and from that moment on, deformation takes place in what will eventually become the product's wall.

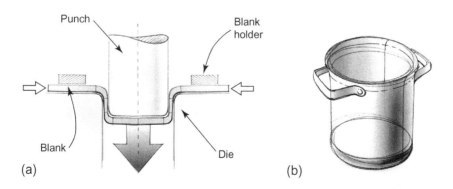

**FIGURE 4.14**

Schematic for (a) deep drawing and (b) typical application.

This causes strain hardening in the wall, so soon after, the not-yet-strain-hardened flange becomes the weakest spot, and further deformation will therefore come from the flange, which now gets drawn inward. Depending on the punch radius and the amount of lubrication, material from the product's bottom may be drawn out, further increasing the deformation potential (as discussed in Section 4.8). The whole process becomes much easier if the material has a large difference between yield stress and tensile strength, and indeed, so-called deep-drawing steels are designed and manufactured to have exactly that property.

---

**EXERCISE 4.37**

Revisit Exercise 4.2 one last time. What is your answer now?

---

## Matched die forming

This method is ideal for mass production, and is the process of choice for the car industry, where cycle times of mere seconds per part are quite normal. It resembles deep drawing and is in fact often (but erroneously) referred to as such, but now the lower side of the blank is fully supported by the lower die (Figure 4.15). In practice, parts can contain stretched ($\varepsilon_1$ and $\varepsilon_2$ both positive) as well as drawn ($\varepsilon_1$ positive, $\varepsilon_2$ negative) and bent areas. Ideally, both dies match perfectly (hence, of course, the method's name), even after taking into account the thickness changes that the blank undergoes. The setup work needed to obtain this match is time-consuming and costly, not to mention that the dies themselves are usually made from tool steel, which is expensive to shape to begin with. The car industry demands extremely high surface quality, again adding cost. Furthermore, most parts require a succession of forming steps—sometimes up to seven operations, including edge trimming—which makes the part-specific investments even higher. Not surprisingly, even large car manufacturers try to share

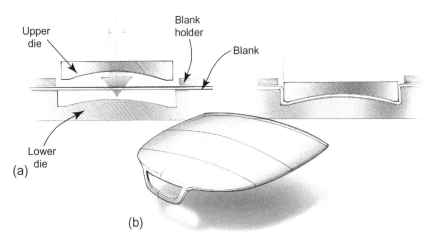

**FIGURE 4.15**

Schematic for (a) matched die forming and (b) typical application.

matched die-formed parts among as many different car "platforms" as possible. However, if somewhat softer alloys are formed and if the requirements for tolerances and finish are less stringent, then matched die forming can be quite attractive even in volumes of just a few thousand units, thanks to "soft tooling" materials such as aluminum, epoxy, or even concrete.

## Sequential die forming

A notable variant of matched die forming is sequential die forming. In this method, a long, thin strip of metal is pulled through a series of presses, each executing one forming step. Each "station" has its own dies, and all dies move up and down at the same time. Only at the last station is the part cut from the strip, so transport between stations is easy: the strip is simply pulled along one station at a time. Sequential die forming is well suited for very large production volumes of small parts. Lamp fittings and shaver head caps are made by sequential die forming, but there are many more examples. The method can also be used for more modest volumes.

## Rubber forming

Originally developed for the aircraft industry, the rubber forming process has moved into many other product categories. In this method, the blank is pressed onto a die by means of a large rubber block. It is similar to matched die forming, but with just one die. Obviously, the costly die matching procedure is eliminated, as the rubber will automatically conform to the shape. Once matched, the rubber is constrained and then has a much greater stiffness. Consequently, the part-specific investments are much smaller and the process can be used for smaller production volumes. On the downside, the method is slower and its form freedom is somewhat less, restricting application to comparatively shallow products.

## Hydroforming

This method can be used for sheets, but it is mainly applied to shape thin-walled tubes and profiles, and this is the version we describe here. In hydroforming, a tube, profile or similar closed semi-finished product is clamped inside a two-piece steel die and connected on both ends to a water (or oil) pressure system. The water is pressurized (commonly to 1,000 bars!), causing the tube to expand like a balloon and assume the inner shape of the die. The main benefits are the homogeneous deformation, even for complex shapes with fine details, and the very close tolerances; it is also quite fast. The method is reserved for high-volume production, as setup time and investment costs are considerable. Audi uses hydroforming to shape pre-bent aluminum extrusion profiles into the A8 A-pillar and roof rail. As for process limitations, consider that the stress level $\sigma$ in a cylindrical tube with radius $R$, thickness $t$, and internal pressure $p$ is given as $\sigma = (p\,R)/t$. So even using very high pressures, alloy strength and tube thickness are eventually limited.

---

**EXERCISE 4.38**

List the six methods discussed in ascending order of design freedom, using a three-point scale for part size (small-medium-large), range of metals (soft-medium-hard), and shape complexity (lower-medium-high).

---

**EXERCISE 4.39**

Now list the six by investment level and finally make a sketch plot of design freedom versus investment. What is your conclusion?

---

**EXERCISE 4.40**

Make two clear, specific guidelines for the design of double-curved parts, aiming to achieve a good trade-off among form freedom, cost (e.g., investments, but also cycle time), and quality (e.g., tolerances, surface finish). Explain by referring to Sections 4.6 to 4.9.

---

**EXERCISE 4.41**

Take another look at your sample products (Exercise 4.1). How can you see your design guidelines reflected in them?

---

## 4.11 **Further Reading**

Sheet metal forming is a world in itself, and especially in the automotive industry it has reached a staggering level of refinement. This chapter provides designers with an introductory basis. Additional relevant concepts to investigate for a deeper understanding include the strain hardening exponent (which essentially captures the stress-strain response beyond the yield stress), the $R$-value (which characterizes anisotropy), and the deep draw ratio (which is a practical measure of formability).

The field also hides many surprises: for instance, why do austenitic stainless steels, which are themselves non-magnetic, suddenly become magnetizable during forming? And what about *twinning-induced plasticity* (TWIP) steels, which combine a yield strength of more than 600 MPa with a strain to failure of 50%? Then there are the numerous exotic manufacturing methods, such as superplastic forming, spinning (comparable to panel-beating), hot matched die forming of martensitic steels (applied in the VW Golf mark VII), and even "explosive forming." The interested reader is referred to the literature listed below, as well as to industrial practice: some things have to be seen to be believed.

Sheet forming does not have a dedicated academic journal, but there is plenty of coverage (along with that for other processes) in, for example, the *International Journal of Materials Forming*.

Professional conferences include the bi-annual International Conference on Sheet Metal (see www .shemet.org), the Automotive Circle for automotive body forming (see www.automotive-circle.com/ Conferences), and the International Deep Drawing Research Group meetings (see www.iddrg.com) that specifically focus on deep drawing and matched die forming.

Ashby, M.F., Shercliff, H.R., Cebon, D., 2013. Materials: Engineering, Science, Processing and Design, third ed. Butterworth-Heinemann, Oxford, UK (Chapters 4 and 6). (*An introductory text at a complementary level to this book, providing detail on material properties and processing in a design-led context. Chapters 4 and 6 include stress-strain behavior and the physical origins of Young's modulus and yield strength.*)

ASM Handbook, Volume 14B: Metalworking: Sheet Forming, 2006 ASM International, Materials Park, Ohio, United States. (*A comprehensive but expensive work on the subject, but worth the investment for everyone who is serious about sheet metal forming. Note that volume 14A focuses on "bulk forming" (i.e., forging, extrusion)—as in Chapters 5 and 6.*)

# Extrusion of Metals

5

## CHAPTER OUTLINE

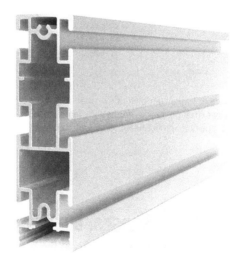

Extrusion is particularly well-suited to aluminum, allowing designers to make custom profiles for modest production volumes. Lower cost is achieved by using standard profiles, such as the one shown here.

*(Image courtesy of Studio Ninaber)*

## 5.1 INTRODUCTION

Extrusion is ideal for making all sorts of metal profiles. Depending on the hardness of the metal that is extruded, the profile cross-section can have a wide variety of shapes and can be solid or hollow, often with multiple internal "chambers." This large form freedom, combined with relatively tight control over the geometrical tolerances, allows us to integrate various functions into the profile, such as flanges, hinges, snap fits, or screw holes. In addition, the extrusion industry offers a comprehensive array of secondary operations, allowing extruded profiles to be bent, machined, and finished with ease, turning the profile into a ready-to-assemble part. What's more, the part-specific investments (i.e., the extrusion dies) can be relatively cheap for simple profiles, typically just 1,000 to 5,000 euros, and times-to-market for extruded products of just six weeks are common. All of this explains the popularity of the process, but it must be added that most of these benefits apply only to aluminum. With copper, brass, or low carbon steel, the possibilities are considerably more limited. Extrusion of other, harder metals is rarer still and limited to basic profile shapes. Magnesium, though relatively soft, can be extruded also, but with more limitations than aluminum. Consequently, around 90% of all metal extrusion is done with aluminum, and most of that comes just from one alloy group: the "6000 series."

This chapter explores the process of extrusion, explaining why certain alloys and shapes are easier to extrude than others. The discussion is limited to "direct hot extrusion," which is by far the most common method. The final section contains some notes on less common extrusion methods. And, of course, this chapter is concerned with metals: the extrusion of plastics (covered in Chapter 13) is a different story altogether.

---

**EXERCISE 5.1**

Look around your house and find one or more examples of extruded metal parts or products. Alternatively, look for products that have not been made by extrusion but—perhaps with some small changes—might have been. Hint: the main characteristic is a prismatic shape (though not necessarily long thin parts, as they may have been sliced from a single extrusion).

---

## 5.2 EXTRUSION AT A GLANCE: PRESSES, BILLETS, AND DIES

At first sight, direct hot extrusion is easy to understand, given the basic layout of the process and the associated terminology (Figure 5.1). At the start of the process, a preheated "billet" is placed inside the press container, which is then pressed by a ram through the extrusion die, and the desired profile comes out. Billets are cylindrical metal bars, made by casting, and are typically around 1 m long. Depending on the ratio between the cross-sectional areas of billet and profile—the *extrusion ratio R*—it is possible to make very long profiles, indeed, longer than can be transported normally. Directly after extrusion, profiles are straightened by stretching and cut to length. The extrusion process and stretching give extruded products slightly different mechanical properties in the longitudinal direction than in the transverse direction. Extrusion is done at elevated temperatures, close to but below the metal's melting temperature—it is a *solid-state deformation process*. Table 5.1 gives some illustrative data for common

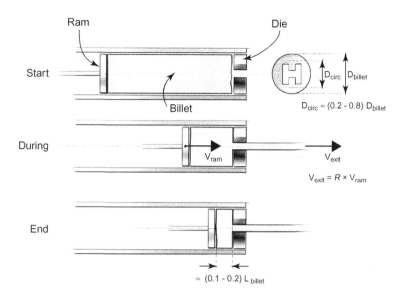

**FIGURE 5.1**

Principle of direct hot extrusion. Note the die diameter and the profile's circumscribed circle.

**Table 5.1** Illustrative Data for Extrusion of Various Extrusion Alloys

|  | Billet Temperature [°C] | Melting Temperature [°C] | Typical Exit Speed [m/min] | Example Alloys |
|---|---|---|---|---|
| "Easy" aluminum alloys | 460-480 | 610-655 | 35-80 | AW-6060* |
| "Moderately difficult" aluminum alloys | 450-500 | 575-650 | 5-30 | AW-6082 |
| "Difficult" aluminum alloys | 420-430 | 475-635 | 0.8-2 | AW-7075 |
| Copper | 780-950 | 1,080 | ~300 | |
| Brass | 550-875 | 952-967 | 60-360 | CuZnAl |
| Carbon steels | 1,100-1,300 | 1,490-1,530 | 1-3[†] | |
| Magnesium alloys | 250-440 | 450-620 | 0.3-30 | |

*AW denotes "aluminum wrought" (i.e., alloys for sheet metal forming, forging, extrusion, and the like).
[†]Estimate—note that carbon steels are very unsuited to extrusion (see also Section 5.6).

extrusion alloys and metals. At these temperatures, metals are considerably softer than at room temperature (RT), but still, extrusion requires high pressures. With billet diameters up to 350 mm, this leads to huge press forces and, consequently, to very heavy machinery. And around 50% of this force is lost to the friction between billet and container (more for smaller billets, less for larger ones), so only half of it is available for the actual deformation.

The billet must be bigger than the profile, which has a circumscribed size defined by the smallest circle that can be drawn around the profile. This is typically 20% to 80% of the billet diameter, depending on alloy type and several profile properties that we will encounter in due course.

Usually, only 80% to 90% of the billet goes through the die in one pass of the ram. The offcut can be recycled as production scrap, and this loss is minimized by placing the next billet right behind the first and continuing the process. This "billet-to-billet extrusion" is only done for non-critical applications, because the "weld seam" joining the two billets is drawn out along the profile and can reduce the strength and ductility. Loading the next billet takes time (for aluminum, extruders typically aim at 80% effective press time), so extrusion is a semi-continuous process.

---

**EXERCISE 5.2**

Billets come in different diameters, usually measured in inches ("). What is the press force range for a 4", 6", or 10" billet, if the pressure on the billet must be 25 to 100 MPa?

---

**EXERCISE 5.3**

In extrusion of relatively hard alloys, the press force often becomes the limiting factor. Explain why it then helps to use shorter billets. What is the consequence in terms of production scrap and, hence, profile costs?

---

**EXERCISE 5.4**

Billets are sometimes pre-heated with a "temperature taper" such that the end near the die is hotter than the other end, near the ram. Explain why. Hint: friction and plastic deformation create heat!

---

## 5.3  A CLOSER LOOK: STRESSES AND STRAINS IN EXTRUSION

If we take a closer look at the stress situation, we first notice that almost the entire billet is under hydrostatic pressure during extrusion: the metal is compressed while being enclosed on all sides, and consequently there is a compressive stress in all three principal directions. The pressure (in $N/m^2$) exerted by the ram *on* the metal is then equal to the stress level *in* the metal (also in $N/m^2$), in any direction. In such a triaxial stress state, metals can withstand considerably higher stresses before yielding than in a uniaxial stress state.

Right at the die exit, the stress state is different: here, the metal is unsupported on one side (i.e., the profile shape) and enclosed on all others. In this state it is weaker than under hydrostatic pressure. Consequently, the deformation of the billet during extrusion gets concentrated in a volume shaped like

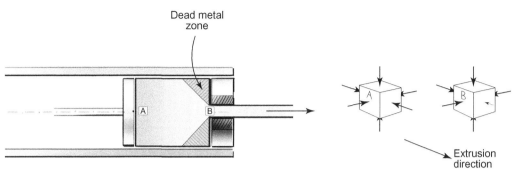

**FIGURE 5.2**

Detail of extrusion, with dead metal zone.

a truncated cone, with the die exit forming the cone tip (Figure 5.2). This also means there is a volume of material that does not deform at all: the "dead metal zone." This zone forms much of the 10% to 20% production scrap mentioned in the previous section.

---

**EXERCISE 5.5**

If the profile circumscribed diameter is small relative to the billet diameter, do we expect the production scrap fraction to be at the low end or the high end of this 10% to 20% range?

---

It is difficult to say exactly how much stress—or pressure—is needed and to relate this to standard material data. For one thing, the actual stress situation near the die is very different from the uniaxial state that applies to standard tensile testing. Linking them would require a suitable flow criterion, but that takes us beyond the scope of this chapter. Also, at extrusion temperatures, metals are considerably softer than at RT, due to the vastly increased mobility of dislocations at higher temperature. Strain hardening mechanisms continue, with more dislocations being generated, but they are now so mobile that they partially eliminate each other (by the mechanism known as "dynamic recovery"). In aluminum this leads to a steady-state dislocation density and reduced flow stress (with the added complexity that it varies with both temperature and strain rate). Certain extrusion alloys even exhibit "dynamic recrystallization," in which the metal's grains are continuously being rebuilt during the process, further reducing strength. Tensile testing of metals is rarely done at such temperatures, so we rely on data from high-temperature tests such as "plane strain compression." Calculating back from the extrusion pressures used in industrial practice, we can estimate that the softer metals typically require around 25 MPa of stress during extrusion, and the harder around 100 MPa.

Now, on to the strains. During hot extrusion, the metal can deform up to 400% (nominal strains), depending on the exact profile geometry. This is, of course, much more than the strain-to-failure of even the most ductile metal at RT, in tension—but in common with all hot deformation processes, extrusion takes place in compression (to avoid necking). The very large strains can only be achieved by the elimination of strain hardening, which would otherwise lead to cracking. For more detail on hot deformation of metals, interested readers are referred to the literature recommended in Section 5.6.

---

**EXERCISE 5.6**

Formability increases with temperature, and necessary pressures decrease. Both are advantageous. What disadvantages could there be to increasing the billet temperature?

---

## 5.4 THE EXTRUSION DIAGRAM

The next key to understanding the extrusion process lies in the "extrusion diagram." This is a plot of the maximum extrusion exit speed $V_{exit}$ that can be reached (usually in m/min) against the initial billet temperature $T_b$ in °C, ignoring any billet temperature taper. It is shown schematically in Figure 5.3—the exact diagram depends, among other things, on the kind of alloy and on the complexity of the profile. The diagram shows that the exit speed is constrained by three different lines, which together define a "process window" for extrusion. Naturally, extruders nearly always strive for maximum speed.

The first limit is the one associated with the maximum press force. Observe how it changes with temperature: the hotter the billet, the softer it gets, leading to progressively less force being required for the process (recall Exercise 5.7). Therefore, line 1 runs from bottom-left to top-right in the diagram. The harder the alloy is, the lower this line will be and the slower the extrusion will be—which is only logical (as illustrated by the data for various aluminum alloys in Table 5.1).

The second limit is related to the fact that most extrusion alloys contain certain phases that melt at lower temperatures than the base metal. In the ubiquitous aluminum 6000 alloys, for instance, this is the intermetallic $Mg_2Si$-phase. This so-called incipient melting temperature $T_{in}$ must not be exceeded to prevent such phases from melting, which could cause "hot cracking" (also known as "hot shortness") or even just surface streaks: the profile emerges with numerous, clearly visible surface defects, which is obviously unacceptable. If the exit speed is increased, the deformation heats up the billet faster and the billet temperature must therefore be progressively lower than $T_{in}$ to avoid hot cracking. Hence, this line 2 also runs from bottom-right to top-left in the diagram. Even for a single alloy type, exact values are difficult to give, because $T_{in}$ depends, among other things, on the thermomechanical "history" of the billet (which determines the size and distribution of the intermetallic particles).

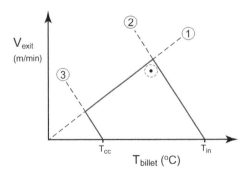

**FIGURE 5.3**

Extrusion diagram. Point denotes optimum combination of speed and temperature. (For color version of this figure, the reader is referred to the online version of this chapter.)

Lines 1 and 2 already define an optimal combination of exit speed and temperature, maximizing production speed. But there is a third line that deserves a mention, associated with "cold cracking." Below a certain temperature $T_{cc}$ extrusion will cause the material to crack: it simply cannot deal with the intense deformation through the die exit. Line 3 runs more or less parallel to line 2, for the same reason: increased speed of deformation increases heating, postponing the danger of cold cracking, though exact data are elusive.

---

**EXERCISE 5.7**

Given a certain alloy and profile shape, which of the three lines on the extrusion diagram do you think are inherent to extrusion as a manufacturing principle, which to the metal in question, and which to the machine that is being used?

---

**EXERCISE 5.8**

Consider the extrusion of two aluminum alloys, one hard, one soft (same press, same profile). Which has the highest possible exit speed? And which has the highest billet temperature at which you would expect it to be extruded? What assumption do you need to make to answer these questions?

---

Remember: what we have said so far about the extrusion diagram is valid for a certain profile complexity and a certain extrusion ratio. The higher this ratio, the more force is required to reach the same extrusion speed. This effect can often be captured in an empirical formula: $p_{ex} = A + B \ln(R)$, where $p_{ex}$ is the extrusion pressure, $A$ and $B$ are material- and profile-dependent parameters, and $R$ is the extrusion ratio. So the slope of line 1 changes with increasing $R$, because the maximum press force becomes more of a limiting factor. But the slopes of lines 2 and 3 also get less, because the amount of deformation is increased, causing the billet to heat up faster and shifting both hot and cold cracking to lower billet temperatures.

---

**EXERCISE 5.9**

Sketch, in a semi-quantitative manner, what happens to the extrusion diagram if the extrusion ratio increases (assume $A/B = 2$). What happens to the maximum extrusion speed?

---

Not surprisingly, the extrusion industry prefers profiles with low extrusion ratios. The best way to do this is by matching the proper extrusion press with a given profile. Therefore, virtually all companies have more than one press, encompassing a range of different sizes. Sometimes extrusion dies are used with multiple profile openings—this is known as multi-strand extrusion—to reduce $R$ (and to increase productivity). It may also happen that profiles are given a higher cross-sectional thickness than they strictly need for design purposes, especially when wide, panel-shaped profiles are involved. This can become costly, as the material price usually is a significant portion of the part price (as a rule of thumb for normal aluminum profiles, this portion is around 50%). Note that extrusion ratios should not get too low either, as some degree of deformation is essential for the process: in practice, for aluminum profiles, $R$ ranges between 20 and 40.

## 5.5 METAL FLOW THROUGH THE DIE: SOLIDS AND HOLLOWS

We now take a closer look at what happens right in the extrusion die. This location is depicted, for a solid Y-shaped profile, in Figure 5.4. The die itself is a flat disk, made of tool steel, with an opening that is the negative of the profile (corrected for known dimensional changes introduced by the process, such as those due to thermal expansion). Notice how the metal does not make contact with the die over the full die thickness, but only on the "bearing surfaces." These crucial surfaces must be shaped (and angled) such that the metal distributes itself evenly over the profile. In this case, near the center of the profile, where there is a concentration of metal flowing through the die, the length of the bearing surface $L_{bear}$ is quite large, whereas at the tree tips, it is considerably smaller. The result is, ideally, a perfectly balanced material flow over the entire cross section of the profile, with all of the metal flowing through at the same speed. The profile will then come out straight. Note that in this example we have a profile with constant cross-section (i.e., the thickness is the same everywhere). In practice—for example, for thicknesses of some 3 mm-bearing lengths typically vary between 3 to 10 mm for ordinary profiles. This is much less than the actual die thickness; however, the remaining material is needed to support the bearing surfaces with sufficient strength and stiffness.

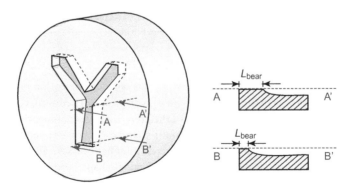

**FIGURE 5.4**

Die for Y-profile "solid," showing how the bearing surface varies over the contour. (For color version of this figure, the reader is referred to the online version of this chapter.)

We just stated "ideally." In practice, ensuring that the profile comes out of the die straight, and within the required tolerances, can be difficult and time-consuming. Despite advancements in computer modeling of the process, extruders still rely largely on trial-and-error plus their own experience in accomplishing this. First, they try to ensure that each side of the profile encounters equal amounts of friction by varying the bearing surface length at each side according to their best judgment. Then they do a first trial run. If the profile comes out bent to one side (and it usually does), they increase the contact angle of the die's bearing surfaces on the opposite side, increasing friction on that side. This process of trial-and-error is repeated until all of the friction forces are balanced and the profile comes out straight.

Another practical consideration is that the profile, upon emerging from the die, is placed on a long table for handling and stretching. The areas of the profile that touch the table can easily be scratched; also, as the material is still quite soft when it comes out, certain details of profiles can even be bent out of true—imagine extruding the letter Y in Figure 5.4: how could we place it on the table without damage?

Changes in thickness are a complicating factor. First off, we must ensure that the thicker sections encounter longer bearing lengths than the thinner ones so that all metal is slowed down to the same speed. Although this is certainly possible (Figure 5.4 showed how), this unfortunately does not solve all our problems. The thinner the metal gets, the more deformation is needed—and, consequently, the more it will heat up locally, causing it to expand relative to the less intensely deformed thicker sections. (The extrusion ratio $R$ applies to the profile as a whole, not necessary to specific locations on that profile.) Upon emerging from the die, those thinner sections will then lose their heat more quickly than the thicker sections, leading to distortion. Unless we somehow manage to balance these effects out carefully, our only refuge is to slow down the process, losing productivity. Not surprisingly, some extruders prefer to avoid such challenges altogether.

---

**EXERCISE 5.12**

What profiles do you think would be easiest to balance out? And what kinds of profile would be more difficult? Hint: consider symmetry (along one or two axes) and differences in wall thickness over the cross-section.

---

The die also has to be sufficiently strong to resist all friction forces. One consequence of this is that for features such as slots in a profile (Figure 5.5), the ratio between slot depth and width is limited. This becomes clear once you visualize the metal flow in three dimensions: the higher this depth-to-width ratio ($H/B$), the higher the shear force on the part of the die—referred to as "the tongue"—that shapes the inside of such slots. If this force gets too high, the tongue will begin to bend to one side, losing tolerances and ultimately failing altogether.

---

**EXERCISE 5.13**

In extrusion of "easy" aluminum alloys, we can typically make slots in profiles with an $H/B$ ratio of up to 3 if the bottom is flat and up to 4 if the bottom is round (see Figure 5.5). How might these ratios change if we use "hard" aluminum alloys, or if we reduce the extrusion speed?

---

As mentioned in the introduction, extrusion allows you to make not just "solids," but also "hollows" (i.e., profiles with one or more cavities), or chambers, in the cross-section. This necessitates the use of two

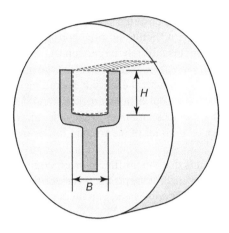

**FIGURE 5.5**

Slot with depth $H$ and width $B$ in a solid profile. Hatched area carries shear force on tongue. (For color version of this figure, the reader is referred to the online version of this chapter.)

or more separate die parts to shape the external and internal contours of the profile. There are three types of solutions, called "bridge dies," "spider dies," and "porthole dies." The last is the most common, with the "mandrel" defining the shape of the internal contour being held in place by "legs" attached to the rear section of the die (Figure 5.6). Note how the first die plate both shapes the outside of the profile and supports the mandrel that protrudes into the second die plate. Porthole dies cost about 140% to 150% of solid-profile dies, for profiles that are otherwise comparable in terms of size, tolerances, and alloy type.

---

**EXERCISE 5.14**

Imagine you wish to extrude the numerals zero to nine. How many numerals have no cavities (i.e., are "solids")? How many have one? How many have more than one?

---

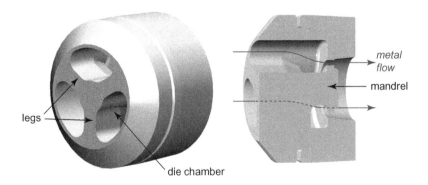

**FIGURE 5.6**

Porthole die for hollow profile. (For color version of this figure, the reader is referred to the online version of this chapter.)

An important consequence of extruding a hollow profile is that the metal flow divides around the legs, leading to longitudinal "weld seams" along the profile where the flows are brought together again to close the hollow shape. (Note that these are quite distinct from the billet-to-billet seams.) To give smooth flow and to minimize the force acting on them, the legs are given a streamlined cross-section in the direction of metal flow. Subtle detailing aims to ensure that the two flows emerge at slightly different speeds to facilitate a sound "friction weld" between the converging metal streams—but not so great as to cause distortion of the profile. Depending on the alloy, these weld seams can be sources of quality problems. They may not be sufficiently ductile, and should be kept away from critical locations in the product. And if the profile is "anodized" later—a common surface treatment to improve the appearance of aluminum parts—then there is a risk that such seams leave ugly lines on the surface. There is considerable freedom in how many bridges to use (at least three, plus at least two per extra cavity beyond the first one) and where to put them, so close communication between designer and extruder on this issue is recommended.

---

**EXERCISE 5.15**

Hollow aluminum profiles offer excellent bending stiffness per unit weight. Can the weld seams become critical spots with respect to bending strength? And what about torsion? Analyze the loading, and conclude where you would *not* want to place these seams.

---

**EXERCISE 5.16**

In dies for solids, slot ratios are limited (see Exercise 5.13). But could we not increase this limit if we were to use a die for hollows? Where would we then place the legs?

---

## 5.6 EXTRUSION METHODS AND MATERIALS

As mentioned in the introduction, our focus is on the most common method of direct hot extrusion. Because the differences between the various extrusion metals are considerable, we structure the discussion here accordingly.

### Extrusion of medium strength heat-treatable aluminum alloys (6000 series)

This alloy group is by far the most popular for extrusion. Its alloys are relatively soft during extrusion (enabling high exit speeds), but they can also be easily heat treated afterwards to gain considerable strength. The heat treatment in question is known as *age hardening*, and it exploits the most effective hardening mechanism in metals, precipitation hardening. It is a three-step process: first, the alloy is "solutionized" at around 530°C to uniformly distribute the alloying elements—particularly magnesium (Mg) and silicon (Si); second, the alloy is rapidly "quenched" to room temperature; and finally, it is "aged" at (typically) 165°C for around 10 hours. In the last step, the Mg and Si precipitate from supersaturated solid solution, to form fine-scale $Mg_2Si$ particles. These precipitates restrict dislocation movement and, hence, increase strength. Steps 1 and 2 of this treatment are commonly integrated within the extrusion process itself—the deformation heating takes the alloy into the solid solution region of the phase diagram (just below the melting point), while the quenching takes place as the profile emerges from the die, often requiring a water spray to ensure a sufficiently rapid cooling rate.

There is a significant quality trade-off here. For a given alloy there is a critical cooling rate that must be exceeded over the whole cross-section to realize its full age hardening potential (and thus target strength). The cooling rate is determined by the shape and size of the profile and the severity of the external quench. But intense water cooling tends to cause greater distortion and warpage, particularly for non-symmetric profiles. So here we have a nice example of complex coupling between the design (profile geometry), the process (temperature history imposed), and the alloy (microstructural response depending on cooling rate). The ease with which a given alloy retains all of the hardening elements in solution is known as its "quench sensitivity." It is also an example of how the final properties can reflect microstructural details inherited from earlier process stages—quench sensitivity is not just dependent on composition but is also affected by the initial casting and "homogenization" heat treatment applied to the billet.

The 6000 series alloys can be extruded in many different sizes, from the tiny iPod nano housing to structural beams for high-speed trains measuring over 60 cm across. In practice, maximum profile widths are limited to those with circumscribed circles up to 25 cm. The possibilities can also be expressed in terms of profile weight, with 0.1 to 20 kg/m being a practical range. Larger profiles are possible, but then profile thickness has to increase also (otherwise $R$ gets too high). Alternatively, larger profiles can usually better be built up out of individual smaller parts, joined together by suitable snap fit joints that are integrated into them, or by longitudinal seam welding (for which friction stir welding is particularly well-suited—see Chapter 12).

---

### EXERCISE 5.17

Look up the strength of the alloys with designation 6060, 6005, and 6082, all in the T6 ("precipitation hardened") temper. Also, look up the percentages of Mg, Si, and manganese (Mn) in these three alloys. Sketch a graph of strength against the total content of alloying elements, and explain the trend you see.

---

### EXERCISE 5.18

Which of these three alloys can be extruded at the highest speed, given a certain press and profile? Explain. Hint: alloying elements also contribute to the strength when they are distributed homogeneously at temperature by solid solution hardening.

---

### EXERCISE 5.19

Which of these three alloys gives the most form freedom, given a certain press and a certain minimum allowable extrusion speed? Explain, using the theory covered in Sections 5.2 to 5.4.

---

**EXERCISE 5.20**

Now give three rules for designing successful profiles, using alloys from the 6000 series. Explain on the basis of the preceding theory or practical considerations. Hint: do not forget to look beyond the issues raised by Exercises 5.17 to 5.19.

---

**EXERCISE 5.21**

How can you make a snap-fit joint between two profiles? And a hinge? And how would you integrate a screw hole into a profile, both parallel to the direction of extrusion and perpendicular to it? Be creative!

---

## Extrusion of high-strength heat-treatable aluminum alloys (7000 series)

For higher strength applications, we switch to the 7000 series, with magnesium, zinc, and usually some copper as the main alloying elements. Its alloys can be precipitation hardened and the heat treatment is integrated into the extrusion process, much as for 6000 series, but with a lower aging temperature (typically 120°C). Quench sensitivity is also a significant factor in these alloys. Alloys from this series offer the highest strengths of all aluminum alloys, of order 25% stronger than alloy 6082. This choice is driven by the goal of reducing product weight, so 7000-series alloys are popular for performance bicycle frames or aircraft parts. Corrosion resistance of such alloys is, however, not as good as those of the 6000 series, and because of the high alloying element content, extrusion speed is lower (see Table 5.1), with obvious cost consequences. Likewise, form freedom is more limited as compared to the previous alloy group.

---

**EXERCISE 5.22**

Recall that as a rule of thumb, 50% of the price of parts made of aluminum profiles is taken up by the price of the aluminum. Think of parts where this ratio is lower and where it is higher. Explain.

---

## Extrusion of other metals

Copper, brass, low carbon steels, magnesium, and a few other metals are also extruded, but much less than aluminum alloys. Here are a few of the applications, along with the main reasons why there are not that many more:

- *Copper alloys*. Main applications are tubing and wiring. Copper is difficult to extrude because of the relatively high temperatures that are required (although exit speeds can be very high, as pure copper contains no phases that can cause hot shortness). Extrusion of copper gets strong competition from several other processes, notably drawing.
- *Brasses*. Mainly applied in heat exchanger tubes and feedstock for forgings. Extrusion of brass has the comparable difficulties and competing processes as copper.

- *Carbon steels.* A few niche applications, such as small-bore tubing and semi-finished products that are difficult to make with other processes. Extrusion of steel is difficult: it requires high temperatures—with the dies themselves also being made of tool steel and with molten glass (!) as a lubricant to avoid excessive die wear. It gets strong competition from roll forming and hot rolling.
- *Magnesium alloys.* Very few applications yet. Extrusion of magnesium would be well-suited for lightweight profiles (for bike frames, ladders, etc.) but is still in its infancy. Hot shortness is a major problem, keeping extrusion speeds disappointingly low.

---

**EXERCISE 5.23**

A "performance index" for minimizing product weight for a given stiffness in bending is $\sqrt{E}/\rho$. Compare the values of this index for aluminum and magnesium. Based on this criterion, how much lighter can a magnesium product be, in theory, if bending stiffness is the design driver? How do both of these materials compare on this criterion with low carbon steel?

---

## 5.7 Further Reading

This chapter has addressed the key features of extrusion as a principle, offering the usual blend of physical constraints and practical possibilities. Useful as this knowledge is for designers, it is equally important to be aware that the extrusion industry offers a wide range of post-extrusion processes, such as bending and hydroforming (see Chapter 4), but also anodizing and other surface treatments (see Chapter 13). Machining is also among these processes, allowing complex features to be added (e.g., fixtures for automotive bumpers). The ubiquitous Apple desktop and laptop computers are milled to shape on the base of an extruded aluminum slab (see Chapter 7).

Apart from direct extrusion, there are two other extrusion methods for metals you may want to explore: *indirect extrusion* and *hydrostatic extrusion*. In the former, the ram and profile move in opposite directions. It requires a cylindrical ram, with the profile emerging through its center. The method is used exclusively for very hard materials (e.g., alloy 7075) and is limited to simple, solid profiles. Hydrostatic extrusion uses some kind of fluid between billet, container, and ram. The pressures are immense (indeed, the fluid can get compressed by some 10%), but the friction between billet and container is virtually eliminated, allowing relatively easy extrusion of "difficult" alloys. The method offers great potential, particularly for magnesium, but it must be said that it is still in the experimental stage. Innovations in extrusion alloys have not been discussed either (e.g., the development of the extra-high-strength AW-6069 alloy). Beyond the process practicalities, there is a lot more to discover about the material science behind the process. The following sources provide a starting point for further understanding.

The extrusion industry itself is now fully mature. Trends are not so much aimed at radically new materials or processes but more at the consolidation of market shares and the refinement of capabilities. More control over tolerances and other quality aspects, as well as on cost aspects such as time-to-market ("right first time") and environmental impacts, are what determine the current industrial agenda.

Davis, J.R., et al. (Eds.), ASM Specialty Handbook. 1993. In: Aluminum and Aluminum Alloys, ASM International, Materials Park, Ohio, United States. (*In-depth material, recommended for anyone who really wants to know what aluminum extrusion is about.*)

European Aluminium Association, 2003–2006. aluMATTER. http://aluminium.matter.org.uk. (*Extensive educational resource on all aspects of aluminum technology—well worth a visit and extended browsing.*)

SAPA design guide. (*This handy book by the Swedish extrusion company SAPA gives many practical design possibilities with extruded aluminum products. Well recommended! It can be ordered via www.sapagroup.com.*)

# Forging of Metals

# 6

Tools have to be strong and wear-resistant, and must have accurate dimensions. This makes forging the ideal manufacturing process. Shown here is a spanner made from high-strength, low-alloy steel, chromium-plated for corrosion resistance.

*(Image courtesy of Studio Ninaber)*

## 6.1 INTRODUCTION

Like casting, forging is an age-old manufacturing principle that over the centuries has reached a high degree of refinement. It is typically used to manufacture metal products that have to withstand very high forces, such as steel spanners, knives, and so on, as well as numerous car components, notably con-rods and suspension arms. If you own a quality bicycle, the chances are it will have forged aluminum cranks and several smaller forgings in its brakes.

Forging takes in a slab, rod, or block of semi-finished solid material and transforms it into the desired shape in a number of blows (often more than one). So the principle uses compressive forces to plastically deform the metal. Like castings, forged parts usually require additional machining: accurate holes will need drilling, flat mating surfaces will need planing, and so on. Compared to casting, forging offers less form freedom but considerably better material properties. The composition of wrought alloys is not constrained by the need for shape castability—wrought metals are in fact less heavily alloyed, but deliver superior strength, ductility and toughness. This is because deformation processing (and associated heat treatments) lead to fine-scale microstructures (from grains down to hardening second phases). The forming process itself contributes significantly to the microstructure evolution, making it challenging to produce uniform properties throughout the part. An example of how the grain structure varies with shaping process is shown in Figure 6.1—the "fibrous" fine-scale structure of the forged part offers a clear advantage over the cast and machined variants. Another benefit of forging is its high speed, making it applicable to very large production volumes. For instance, threaded fasteners are also forged by a special method known as *cold heading*, whereas coins are produced by *coining*; in both cases, annual production volumes can run into the hundreds of millions!

At first sight, forging is a highly versatile process, delivering products that range widely in weight from grams to tons, and can be made from nearly any metal. Parts can be purely functional and produced automatically by the thousands, or they can be delicately sculptured "one off" objects made by hand by skilled artists. There are numerous different forging methods to choose from, and forging equipment comes in all shapes and sizes, the largest applying huge forces through mechanical or hydraulic means. But the manufacturing triangle of forging can be easily explored once we know some basic terminology and have a first, simple model of the process.

---

**EXERCISE 6.1**

Look around your house and locate one or more forged products. Hint: don't forget to inspect your bike, or to explore a toolbox.

---

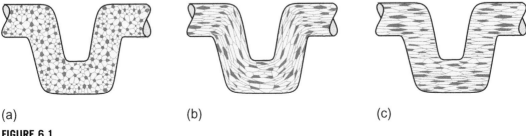

(a)                              (b)                              (c)

**FIGURE 6.1**

Schematic grain structures in components that were: (a) cast, (b) forged, and (c) machined. (For color version of this figure, the reader is referred to the online version of this chapter.)

## 6.2 FORGING AT A GLANCE: BASIC TERMINOLOGY AND PROCESS MODELING

In the simplest of all forging operations, a short cylinder of metal—referred to as a *billet, workpiece, blank,* or *slug*—is compressed between two flat plates, or *dies,* so that its height decreases (Figure 6.2). Because the volume remains constant during plastic deformation, the billet's diameter increases in the process. This particular deformation mode is called "upsetting," one of several standard variants that are documented in forging handbooks.

It is instructive to observe how the required force increases with decreasing billet height (i.e., against die displacement). This dependency is shown in Figure 6.3 (solid line). After some initial bedding in, mainly to do with leveling out surface irregularities, we see an exponential relationship: it

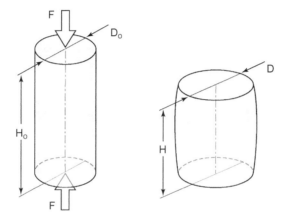

**FIGURE 6.2**

Upsetting of a cylindrical billet. (For color version of this figure, the reader is referred to the online version of this chapter.)

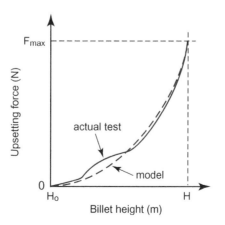

**FIGURE 6.3**

Force-displacement curve during upsetting. (For color version of this figure, the reader is referred to the online version of this chapter.)

requires progressively more force to reduce height. This leads us directly to an important insight into forging, which is that large shape changes in forging can require large forces. Another observation we can readily make during actual upsetting tests is that the billet does not remain cylindrical but instead assumes a barrel-like shape.

There are three reasons why the force rises so rapidly during upsetting. The first is that the billet's cross-section increases. So even if the yield stress required for deformation would be constant, the force would increase. But if we deform the metal cold, the yield stress increases continuously because of strain hardening (also called work hardening). Forging can also be done "warm" and "hot" to lower the yield stress (and thus forging load), but for now we assume cold forging, where strain hardening takes place. The third reason is more subtle and is due to the *friction* between the two dies and the billet. Friction always plays a role in forging, even with the lubrication that is usually applied. Friction causes the stress state in the billet to become multi-axial. A full analysis is beyond the scope of this book, but in essence the material becomes harder to yield when compressive stresses act in two or three perpendicular directions. Depending on the level of friction and the height:width ratio of the billet, the average compressive stress can in fact be as high as 5 to 10 times the uniaxial yield stress.

Can we make a simple model of force against displacement during upsetting? It turns out that we can, at least to account for the first two effects of billet geometry and strain hardening. We therefore make the following assumptions:

1. Friction is ignored and hence the compression is uniaxial (so no barreling occurs).
2. The billet is short enough that it does not buckle, but it deforms under pure compression (typically $H \leq 4D$).
3. The dies are perfectly rigid and do not yield.

For large plastic strains, it makes most sense to capture the material response using *true stress* $\sigma_t$ and *true strain* $\varepsilon_t$ (the latter was introduced in Chapter 4). The true stress is the force $F$ divided by the actual cross-sectional area, which increases during upsetting (i.e., not divided by the original area, which would give us the nominal stress). The true strain also accounts for the continuous change in the billet length with deformation and is defined as $\varepsilon_t = \ln(H/H_o)$.

For many alloys, the relationship between $\sigma_t$ and $\varepsilon_t$ can be fitted to an equation of the form

$$\sigma_t = K \varepsilon_t{}^n \quad \text{or} \quad \log(\sigma_t) = \log(K \varepsilon_t{}^n) = \log(K) + n \log(\varepsilon_t) \tag{6.1}$$

The parameter $n$ is called the strain hardening exponent, and $K$ is a material constant. Figure 6.4 shows the shape of this true stress-strain relationship—on log scales (to the right) the exponent $n$ is equal to the slope of the line. Table 6.1 provides typical data for $K$ and $n$ for several common forging metals.

---

**EXERCISE 6.2**

Using the data in Table 6.1, draw the true compressive stress-strain curves of these four metals (on a log-log scale).

---

**EXERCISE 6.3**

Use the data for $K$ and $n$ to estimate a compressive yield stress for each alloy (assuming the yield stress is defined as the stress at which the plastic strain is equal to 0.2%).

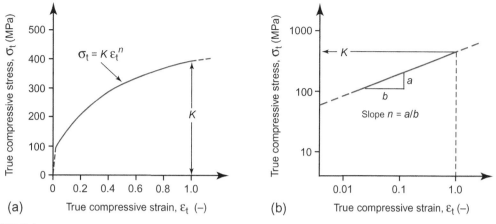

**FIGURE 6.4**

True stress-strain curves (compression): (a) linear scales, (b) log scales. (For color version of this figure, the reader is referred to the online version of this chapter.)

**Table 6.1** Key Data for Several Forging Metals

| Metal | K (MPa) | n |
|---|---|---|
| High-carbon, low-alloy steel (34 Cr 4) | 870-970 | 0.13 |
| Stainless steel AISI 304 | 1300-1500 | 0.42 |
| Aluminum 6082–T4 | 210-260 | 0.20 |
| Brass (CuZn37) | 710-880 | 0.46 |

With this model, we can now derive a formula for the force $F_{up}$ required for upsetting a cylindrical billet with initial radius $R_o$ from an initial height $H_o$ to a final height $H$ (noting that volume is conserved: $\pi R_o^2 H_o = \pi R^2 H$):

$$F_{up} = \pi R^2 K \varepsilon_t^n = \pi R_o^2 (H_o/H) K \left[ \ln (H/H_o) \right]^n \tag{6.2}$$

This is plotted in Figure 6.3 as the dotted line. It predicts the observations well, if friction is kept low. Upsetting tests are routinely used to determine the properties of $K$ and $n$ for a given alloy.

---

**EXERCISE 6.4**

Choose a metal from Table 6.1. What are the forces required to upset a billet of this metal by 25% and 50% (neglecting the influence of friction)?

---

Under such forces, the billet deforms plastically, but the dies will also change shape, albeit elastically. Considering that dies are made of tool steels, the importance of die deformation depends on the relative stiffness of the billet material compared to tool steel and on the severity of the shape change imposed (i.e., the peak compressive stress applied). This has one peculiar effect on tolerances that all designers should know. In a typical upsetting-type forging operation, it is more difficult to obtain close tolerances along the main axis, where the tool deformation is highest, than along the two perpendicular axes.

---

**EXERCISE 6.5**

Which metals were used for your sample products (see Exercise 6.1)? Which parts of these products require close tolerances to function? For instance, for a typical spanner that would be the distance between the two beaks, as well as how parallel they are.

---

Upsetting is only one of the possible deformation modes—another common one is "drawing out," in which we increase the billet length while reducing its cross-section. But the take-home message is that cold forging involves massive forces, especially for strong metals and substantial shape changes. This is part of the reason why forging is often done with successive blows, gradually changing the billet's shape, sometimes with a series of dies that redistribute the material progressively to the final product shape. Another difficulty is the limited deformation potential during cold forging: even though the loading is compressive, the imposed strains are still tensile in some directions and may exceed the metal's capacity to deform without tearing. It is to overcome these limitations that forging is often—for some metals, exclusively—done at elevated temperatures, to which we therefore turn our attention now.

---

## 6.3 COLD, WARM, AND HOT FORGING: BENEFITS AND DRAWBACKS

Cold forging (indeed, all cold forming) is generally defined as forging at temperatures below $0.4\ T_{m}$, with $T_{m}$ being the metal's melting temperature in Kelvin. Observe immediately that "cold" can still be hundreds of degrees Celsius. Cold rolling of steel, for instance, is typically done at around $400°C$, when its formability is considerably better than at room temperature (RT). Similarly, "hot forging" is defined as forging at temperatures above $0.8\ T_{m}$. Again, "hot" is relative: for pure tin, for instance, this higher limit is just $130°C$. Forging at temperatures between these two limits shows a gradual transition between the two regimes and may be referred to as "warm forging." Note that with few exceptions, such as pure aluminum or eutectic compositions, metallic alloys have a solidification range, so in such cases we use the lower limit of this range, the solidus temperature $T_{solidus}$, for $T_{m}$.

---

**EXERCISE 6.6**

Determine the "cold," "warm," and "hot" temperature ranges, in degrees Celsius, for high-carbon steel and compare with those for one or two metals with a lower $T_{m}$ (such as Al, Mg, or Zn).

---

Above $0.8\ T_{m}$, metals have a much lower yield strength than at room temperature. Consequently, the forces necessary to produce the deformation are significantly lowered. (Recall from the previous chapter that these are the same reasons why extrusion of metals is also done hot.) Importantly, however, the mechanism of yielding changes to *creep*, in which dislocation motion is modified by diffusion.

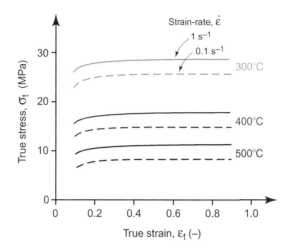

**FIGURE 6.5**

Typical hot deformation data for aluminum alloy AW-6082 (compression). (For color version of this figure, the reader is referred to the online version of this chapter.)

In some alloys, such as aluminum, *dynamic recovery* occurs—dislocations rearrange and annihilate as fast as they are produced by work hardening, so that a steady-state flow stress is reached, with no increase in yield stress with strain. In carbon steels, *dynamic recrystallization* occurs, with the whole grain structure being continuously renewed during deformation, again wiping out work hardening—in this case the yield stress can actually peak and then decrease with strain. Not only is the yield stress reduced in hot deformation, but the ductility is increased, allowing greater design freedom. But a complicating factor is that the yield stress now becomes dependent on the rate of deformation. Figure 6.5 shows data for a common wrought aluminum alloy, for temperatures and strain-rates typical of forging. A different formula for yield stress is commonly fitted to data of this sort, linking the steady-state stress $\sigma$ to temperature $T$ and strain-rate $\dot{\varepsilon}$:

$$Z = Z_o \left[ \sinh \left( \sigma / \sigma_o \right) \right]^n \quad \text{with} \quad Z = \dot{\varepsilon} \exp \left( Q/RT \right) \tag{6.3}$$

where $Z$ is the "Zener-Hollomon parameter" (in $s^{-1}$), $Z_o$ and $\sigma_o$ are material-dependent constants (also in $s^{-1}$ and MPa, respectively), $Q$ is an "activation energy" for hot deformation (in J/mol), and $R$ is the universal gas constant ($= 8.314$ J/mol.K). Note that there is no longer any dependence on the amount of strain, just the strain-rate.

**EXERCISE 6.7**

An aluminum forging process is conducted at 385°C. To improve productivity, we wish to increase the strain-rate by a factor of 10 but cannot increase the forging load (i.e., applied stress). By how much should the operating temperature be raised to achieve this? For hot deformation of aluminum in this regime, the activation energy $Q = 156$ kJ/mol. (Hint: for the same yield stress, the value of $Z$ will remain unchanged.)

**EXERCISE 6.8**

Compare the range of hot yield stress for the 6000-series aluminum alloy in Figure 6.5 with the data for 6082-T6 in Chapter 4. By roughly what factor is the material strength reduced?

If hot forging only had benefits it would be the norm, but it has four main drawbacks. The first is the reduced control over tolerances. In cooling down, hot forgings can distort and warp, or build up internal stresses, due to differential cooling. For parts that receive secondary machining afterwards, this need not be a problem, but it remains a quality issue. Second, oxidation can be severe at high temperatures, leading to a loss of material as this cracks and spalls from the surface, affecting both tolerance and surface finish. Third, there is no increase in strength due to the absence of strain hardening, so all else being equal, parts made by hot forging are less strong than cold-forged parts. The high temperatures often lead to recrystallization and grain growth, effectively softening the material instead of strengthening it. Finally, there is the increased cost, to supply the energy required for heating, and the facilities for heating and handling large numbers of hot items safely. Finding heat-resistant lubricants is also not easy (recall how extrusion of low carbon steel requires liquid glass as a lubricant). These costs are partially offset by the use of less powerful and, hence, cheaper equipment, but in general hot forging is more expensive than its cold counterpart.

**EXERCISE 6.9**

Hot forging can in principle be done with dies that are "cold" or "hot" (i.e., heated to the same temperature as the billet). Both have benefits and drawbacks. Given that the dies are made of tool steel, which options are applicable to forging of aluminum? And which for forging of steel?

## 6.4 DIGGING DEEPER: FRICTION, FLASH, AND MULTI-STEP FORGING

We return to upsetting of a cylindrical billet, as shown earlier in Section 6.2. If we take a closer look (Figure 6.6), we observe that it consists of two different zones, with rather different states and degrees of deformation—although in reality these merge into one another seamlessly. The first is the core material, or "dead-metal zone" in which the deformation is close to zero, or well below the nominal overall strain (directly analogous to that observed in extrusion; see Chapter 5). For the diameter to increase, the billet must slide over the die, but this is resisted by friction (acting radially inward). Equilibrium of the stresses dictates that at the center there are compressive stresses acting horizontally, as well as vertically. These stresses fall gradually to zero at the free surface. The triaxial stress state makes the material more resistant to deformation (plasticity theory tells us that yielding is governed by the *differences* between the normal stresses acting in perpendicular directions). To accommodate the overall axial shortening, the plastic deformation is therefore concentrated into the remainder of the billet. Deformation becomes progressively easier with increasing distance from the dies, and volume must be conserved, giving the "barreling" observed. Apart from the overall loss of shape, the lack of homogeneity in plastic deformation can lead to a significant variation in microstructure (and thus properties) over the volume.

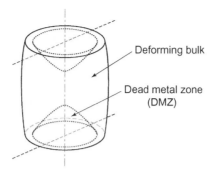

**FIGURE 6.6**

Inhomogeneous deformation of billet during upsetting, due to friction, leading to a "dead metal zone" (DMZ). (For color version of this figure, the reader is referred to the online version of this chapter.)

In cold deformation, non-uniform strain leads to non-uniform hardening. And if we deform warm or hot, there will be also be variations in strain-rate and temperature—the latter due to heat losses to the dies and the surrounding air, as well as heat generation due to the plastic work. Even if full recrystallization takes place subsequently, there will be variations in the resulting grain size (which also depends on strain, strain-rate, and temperature). To make matters worse, and a particular issue for warm forging, there may be regions of recrystallized and unrecrystallized material within the same forging.

In practice, this type of "open-die forging" between parallel platens is limited to simple shape changes of rectangular and cylindrical billets, usually as the precursor to subsequent "closed-die forging" in which a fully three-dimensional shape is formed (more details in Section 6.6). Figure 6.7 shows the forging of a cogwheel, comprising cylindrical sections of different diameters. The cavity in the die defines the part geometry, and the goal is obviously to fill this completely—forging using profiled dies is known as "impression die forging." But there is some freedom to choose the size and shape of the starting billet—this is optimized as far as possible to keep the loads down and to give smooth continuous deformation into the die cavity. Wherever contact is made with the die, friction will restrict the flow over the die surface, and the load rises, while the contacts constrain and direct the flow of material into the cavities. Poor process design can lead to incomplete filling of the die or defects such as "cold shuts," where material is folded over on itself, trapping a layer of surface oxide and leaving a plane of weakness—a possible origin for a fatigue crack or fracture in service.

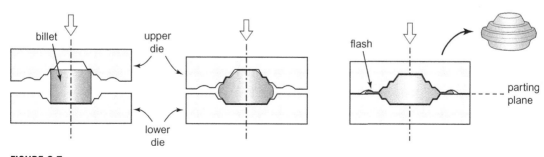

**FIGURE 6.7**

Impression die forging of a cogwheel in three stages. (For color version of this figure, the reader is referred to the online version of this chapter.)

In impression die forging it is usual to use an excess volume of material. The dies cannot then close fully, and a thin layer of metal known as a "flash" is squeezed between them at the edge of the forging (see the third stage of Figure 6.7). Typically 10% to 20% of the billet ends up in the flash. One benefit of producing a flash is that a lower degree of precision is needed in cutting the original billet to size, and it can easily be trimmed off afterward and recycled as production scrap, as simple secondary operations. A more subtle benefit of flash formation is that the pressure on the material here is very high (due to the friction acting on material that is much thinner than its width). This promotes full take-up of the profile in the die—material flows more easily into the last corners and fine features than it does into the flash region. Proper process design generates a ring of flash surrounding the forging on the die parting line. Impression die forging has the added benefit that as the die is filled, a compressive stress acts radially inward, and there is no further tensile strain in the circumferential direction, preventing surface cracking.

### EXERCISE 6.10

Observe your sample products (see Exercise 6.1). Can you see where any flash has been removed?

The simple cogwheel shown in Figure 6.7 requires only modest axisymmetric deformations and can be forged in a single step, even with cold forging. However, if we design a part that involves larger deformations, this may no longer be possible, as the billet's surface may undergo too great a tensile strain before it is constrained by the dies, and starts to crack. Especially for more complex parts, it is therefore common to deploy a sequence of forging steps, with the part gradually obtaining its final shape. Increased form freedom is not the only advantage of this approach: if done well, it can also yield more homogeneous deformation and, hence, more constant material properties. Figure 6.8 shows this

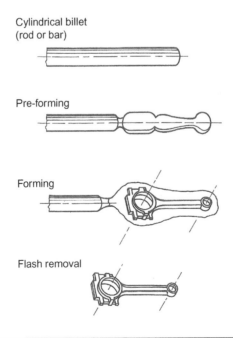

Cylindrical billet
(rod or bar)

Pre-forming

Forming

Flash removal

**FIGURE 6.8**

Multi-step impression-die forging of an automotive con-rod.

for a typical automotive forging: the "con-rod," an essential and highly loaded engine part connecting the piston to the crankshaft. Note how the job of the first "preforming" deformation is to redistribute the material axially to give the appropriate bulk of material locally for big and small ends of the con-rod. Each step requires a separate set of dies so this is a costly setup, but the result can be a homogeneously deformed part, with maximum form freedom. In cold forging with this technique, a major issue to consider in laying out such sequences is that material that is predeformed in the early stages will be strain hardened to some degree, resisting subsequent deformation. Again, hot deformation gives much greater form freedom, and the loss of strain hardening is not a problem if the alloy is chosen to provide an alternative mechanism of hardening. Just as for extrusion, we can exploit *heat treatment* after hot forming to optimize the trade-off between low strength and formability during shaping and high strength in subsequent service. This is such an important characteristic of processing carbon and low alloy steels that we will take a short diversion here to discuss it.

## 6.5 SUPPLEMENTARY TOPIC: HEAT TREATMENT OF STEELS

As noted in the introduction, forging is often applied for strength-critical parts and products. To maximize strength and minimize part weight, and hence cost, it is common to give forgings some kind of heat treatment. The previous chapter touched upon the precipitation hardening of extruded aluminum parts, and indeed, this particular heat treatment is also common for forged aluminum parts: the billet is hot deformed in its soft, "solutionized" state (when strength is minimum and ductility is maximum), then it is quenched and aged to obtain the final "T6" condition and properties. For many plain carbon and low alloy steels, an analogous treatment exists—the "quench and temper" (QT) process. Given the great importance of forged steel in general, and of this ubiquitous heat treatment of forged steel parts in particular, we briefly go into it here. A full explanation would merit its own chapter, covering the iron-carbon phase diagram and much else, but this is beyond the scope of this book. For more detail, see the "Further Reading" section.

---

**EXERCISE 6.11**

Look up the compositions of low, medium, and high carbon steels. What are the values for the carbon content (C) that designate the boundaries between these three groups of steel? Do low alloy and high alloy steels also contain carbon?

---

   The QT treatment hinges on iron's ability to assume different crystal structures. This *allotropy* is uncommon among structural metals: only titanium possesses it also. From RT to 911°C, iron consists of grains with a body-centered cubic (BCC) structure, known as *ferrite* or $\alpha$-iron. Above this temperature, the structure changes to face-centered cubic (FCC), now referred to as *austenite* or $\gamma$-iron. (And, for completeness, it flips back to BCC for a short temperature range just before melting.) The key characteristic of this in steels is that FCC iron dissolves carbon readily (up to around 2%), whereas BCC does not (maximum 0.035%). The basis of the QT treatment is therefore dissolving carbon in the austenite phase, and then cooling and reheating the material to temperatures at which ferrite forms with a second (hard) phase, iron carbide, also known as *cementite*.

   Low carbon steels are not amenable to QT treatments, but are normalized, meaning that they are heated to austenite temperatures and slowly cooled. The same phases form in the microstructure—

ferrite and cementite—but mostly as grains of ferrite, with a small proportion of two-phase grains (with the name "pearlite," thanks to its mother-of-pearl like sheen under the microscope) containing almost all the carbon. This is a reasonably strong but very tough material—the workhorse alloy of construction and steel sheet products.

To obtain the QT response, the carbon content needs to exceed about 0.3%, for both plain carbon and alloy steels. The initial step is the same—austenitization—which for small forgings is typically done in a continuous-throughput oven, under a controlled atmosphere, allowing at least 30 minutes for the full $\alpha$-$\gamma$ transition and dissolution of the carbon into solid solution. Larger objects are processed in batches or piece by piece, and require longer soaking times because of the time it takes for their interior to heat up (steel is a relatively poor conductor among the metals). But instead of cooling slowly, we then *quench* (i.e., rapidly cool) to RT using an oil or saltwater bath. This prevents the formation of cementite, and results in an unusual microstructure called *martensite*. The FCC to BCC transformation still takes place in the iron, as this does not require diffusion, but the BCC lattice is now *supersaturated* in carbon—well above the concentration it normally tolerates. This distorts the BCC structure, giving martensite unusual properties.

With a yield stress of some 2,000 MPa, martensite is incredibly strong, but unfortunately it has negligible toughness, so it is far too brittle for bulk practical use (though it is produced in the form of a thin surface layer for wear resistance). After quenching, the *tempering* step involves bringing the steel to a temperature of around 500° to 600°C, at which temperature the structure transforms into "tempered martensite." The carbon diffuses to form iron carbide but now is dispersed throughout the ferrite on a uniform, fine scale. The resulting strength is below that of martensite, but ductility and toughness are restored, and (crucially) the yield stress is around two to three times higher than the same alloy in the normalized condition. The longer or hotter the tempering treatment, the softer and more ductile the material gets. So the precise tempering treatment depends on the application. Plain high carbon steels can end up with yield strengths between 600 and 900 MPa, with a strain-to-failure of 9% to 12%. Tempering is done in ovens, fluidized beds, or salt baths, the time again depending on component size to allow the interior to heat up. After tempering, the parts are allowed to cool down gradually in ambient air.

---

**EXERCISE 6.12**

What will the ultimate strength be if tempering is too short: too high or too low? And what if the austenitizing step is too short? How important is the factor of time in this respect?

---

A key concept in the QT treatment is that there is a "critical cooling rate" (CCR) that must be exceeded in order to form 100% martensite, which is the essential precursor for the tempering step. The more carbon we add to the steel, the more time we get to quench. But even plain high carbon steels have a high critical cooling rate. The problem is therefore one of a limit on the component size that can be heat treated by QT—however ferociously we quench the outside of a component, the interior cools at a rate which is largely dictated by the (poor) thermal conductivity of steel and the component size.

(There is a close analogy here with the cooling time for castings, captured in Chvorinov's rule; see Chapter 3.) The solution to this dilemma is more metallurgy. The addition of just a few percentages of certain elements (nickel, chromium, and molybdenum being the most common) radically changes the rate at which ferrite and pearlite can form on cooling—opening up much more time to conduct the quench to martensite (i.e., lowering the CCR). This property of a QT steel is called its *hardenability*—its ability to form 100% martensite on quenching. For convenience, it is measured numerically in terms of the maximum diameter of round bar that will achieve this (rather than a cooling rate, because designers are interested in part dimensions).

So for small, thin forgings (which are easy to quench to the core), we may be OK with a plain carbon steel of 0.3% to 0.4% carbon, but for thick, heavy parts we are likely to need a "low alloy steel" (i.e., a medium carbon steel containing up to 8% of alloying additions). High alloy steels, as you might guess, are carbon steels with a higher alloy content (typically 20%, including vanadium, tungsten, and boron). Boron in particular is an effective addition with respect to hardenability: just 30 ppm of this element can have the same effect as 1% nickel. Some high alloy steels may even allow "air quenching" (i.e., cooling with air instead of liquids). This greatly reduces the risk of thermal distortion of parts during quenching and also avoids "quench cracking"—don't forget that the thermal stresses are being applied to a very brittle, martensitic microstructure. A secondary benefit of using certain high alloy steels is that, during tempering, alloy carbides form (instead of some of the iron carbide). These carbides are stable to much higher temperatures, giving useful high temperature strength and wear resistance, for example, in cutting tools.

In passing, note that some highly alloyed steels are not heat-treatable, because they do not contain much (or any) carbon. The most common stainless steels, AISI 304 and 316, contain 18% chromium and 8% nickel, but they are not strictly steels at all, containing next to no carbon (they are strictly "ferro-alloys"). These "austenitic stainless steels" remain austenitic right down to room temperature (and below)—the alloying is not for heat treatment but for corrosion resistance and enhanced toughness at cryogenic temperatures. However, there are "ferritic" and "martensitic" stainless steels, containing 13% or more chromium and no nickel but with a medium to high carbon content, and these *can* be QT treated—useful for producing sharp, wear-resistant cutting edges on cutlery and surgical instruments.

---

**EXERCISE 6.13**

Does the QT treatment also affect the Young's modulus of the steel? Explain your reasoning.

---

The QT treatment has been in use for a long time, with the oldest applications dating back to around 1100 BCE. In those days it must have seemed magical how a single substance could change hardness so strongly. Today, we know why and how it happens, and we understand the influence of the alloying elements on the final properties, also in relation to shape, size, and the time-temperature-deformation history of the forging. Still, the number of possible permutations of the many variables is huge, and not every metallurgical mystery has been solved yet.

The important message therefore is that design with heat-treatable steels is a multi-faceted, coupled problem—the component size influences the cooling and reheating histories, which in turn determine the choice of the appropriate steel to give the target properties, which in turn determine the exact QT process conditions required. Alloy additions are expensive—it is always worth checking whether plain carbon steels will do before moving up to the low alloy steels. Perhaps most important, it is all too easy to set up conflicting requirements for QT treatment of forgings (e.g., between bulk yield stress, surface hardness, ductility, machinability, and costs). So a closer look at the literature is in order for those who want to know more and enter into a fruitful discussion with experts and suppliers. And remember that the QT treatment, although common, is by no means the only heat treatment available for steel forgings. Surface heat treatments are also of interest to the designer: but we will come to these in Chapter 13.

## 6.6 FORGING METHODS

As can be expected for any well-established manufacturing principle, forging is done today using a considerable number of methods. For designers, the main four to study are open-die drop forging, closed-die drop forging, press forging, and net shape forging, all covered here. We will also say a few words about two specific methods for certain products: cold heading and coining.

### Open-die drop forging

In this method, also known simply as "open-die forging," a metal block or rod is manipulated (either manually or with robots) between two dies that move up and down at a fixed rhythm and with a fixed force. Figure 6.9 shows this method being applied to progressively reduce the diameter of a large cylinder, a process known as "cogging" (actually the first step in producing the nickel superalloy blanks for jet engine turbine disks). It is essentially a mechanized version of how blacksmiths hammer workpieces over an anvil, and consequently, the method is also referred to as "smith forging." The dies are usually flat; sometimes they have simple shapes (e.g., V-shaped) to cut off sections of the workpiece. Open-die forging is unsuited for large production volumes because of its low speed, but thanks to its lack of part-specific investments, it is well-suited to small series and even prototyping. Another key application is to give semi-finished products some degree of pre-deformation, taking in—for instance—a cast ingot and

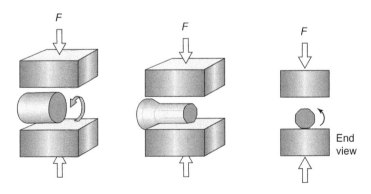

**FIGURE 6.9**

Schematic of the "cogging" process, using open-die drop forging. (For color version of this figure, the reader is referred to the online version of this chapter.)

drawing it out into a longer, thinner block with improved wrought microstructure and material properties. The method can operate at any temperature and can handle large workpieces.

---

**EXERCISE 6.14**

Which are likely to be processed hot with this method: small or large workpieces?

---

## Closed-die drop forging

If a fairly complex part with ordinary tolerances must be forged in large numbers, this method is the one you will probably end up using. Despite the name, it is not strictly a closed-die operation, as the dies do not entirely close off the part, allowing flash formation—it is therefore also referred to as "impression-die forging." Although one-step operations exist, the method usually forms parts in a short sequence of steps, with four being a typical number. Recall the con-rod example (see Figure 6.8): step 1 shows preforming, step 2 the actual part forming, and step 3 the flash removal. As there are multiple die sets, one for each step, the part-specific investments are high (typically somewhat higher than for a high-pressure metal die casting of the same size), but form freedom is good and production speed can be mere seconds for small parts. Closed-die drop forging can be done cold, warm, or hot, and thanks to the development of huge machines, it can also handle an impressive size range—although the usual and obvious dependencies between temperature, size, and type of metal apply as before (i.e., inherent to any forging method).

---

**EXERCISE 6.15**

Which of the following attributes of the part affect die life span: metal type, complexity (i.e., requiring more or less deformation), small radii, and fine detailing, narrow tolerances? So how do these attributes affect part cost price?

---

In common with shaped casting, the part and dies have to be designed so it can be taken out of the dies. This means some design guidelines are similar to casting, for example, with positive draft angles relative to a well-chosen parting plane. The parting plane itself is usually flat, and always perpendicular to the direction of the drop force; furthermore, in a workable die design, the flash must also end up in this plane. Form freedom is less than for casting: for instance, ribs cannot be much higher than they are wide, radii must be very generous to promote metal flow and extend die lifetime, and draft angles of at least 5 degrees (for steel parts) are necessary. Undercuts are not possible, but subsequent machining of parts can easily realize such details; in this light it is interesting to know that most forging companies have considerable in-house facilities for such secondary processing, as well as in-house equipment for heat treatments.

---

**EXERCISE 6.16**

Draw up three design rules for cold closed-die drop forging of low alloy steel parts. How do these rules change if we would have hot forging? And how do they change if instead of steel we would have aluminum?

---

---

**EXERCISE 6.17**

Identify a sample part from those you found in Exercise 6.1 that has been manufactured using closed-die drop forging, and identify the location of the parting plane. What kind of secondary machining has been used, and why? Hint: apart from flash removal, also consider tolerances.

---

## Press forging

A key drawback of drop forging is that it is essentially a "brute-force" process, with limited potential for process control. *Press forging* has been developed as an alternative. In this method, the workpiece is formed between two part-specific dies that are brought together at low speed using hydraulics, with precise control over the force and displacement. Like the previous method, it generates flash. Because press forging offers some control over the strain-rate—which also influences microstructure and final properties—press forging can offer superior quality. Depending on the part in question, it can also be more economical than the previous method, although it is difficult to predict in advance which one will "win." Note that because die contact lasts for seconds rather than milliseconds, any hot workpiece will heat up the dies, and reheating the workpiece halfway through the process may be necessary. In this light it is beneficial that most lubricants act as insulators, limiting heat transfer between die and workpiece.

## Net-shape forging

If we exercise tight control over the billet preparation and if we can predict the tool and die deformation (as well as die wear) very well, then we can, in principle, get the workpiece in exactly the right shape, dispensing with the need for secondary machining. We can then speak of *(near) net-shape forging*. Although it is basically closed-die forging done accurately, it is generally considered to be a separate method. It can also do away with the need to turn part of the workpiece into flash. This is beneficial for more expensive materials (e.g., titanium), because even though flash can be recycled, this still costs time and money and always involves some losses. Another advantage is somewhat increased form freedom (e.g., sharper radii, smaller draft angles). Because of the more costly preparation, net-shape forging is only done when the functional demands and quality constraints are leading, especially if production volumes are limited. Even then, it is rarely used for stronger metals, as it works better with aluminum, for example.

## Cold heading

This specialized method is deployed exclusively for fasteners and fastener-like objects. It is done cold (at RT, in fact) and predominantly with plain carbon and low alloy steels. The input is a coil of thick metal wire coated for lubrication, which is cut to length and fed into the "bolt maker." This machine shapes one end of the billet (i.e., the end that will become the fastener's head) blow by blow, transferring it to the next die between blows. Figure 6.10 shows the various shapes that the billet assumes in the process. The necessary dies—two for simple screws and up to five for complex flange bolts—are joined together into a single block, which can be quickly exchanged once it has worn out. After its head is formed, the fastener's thread is made by rolling the parts between two grooved plates (thread cutting

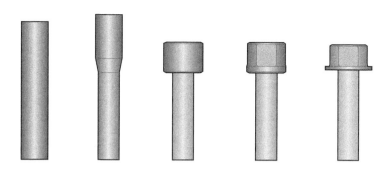

**FIGURE 6.10**

Sequential forming of bolt head by cold heading. (For color version of this figure, the reader is referred to the online version of this chapter.)

would also be possible but would cut through the grain structure and diminish the material properties; it would also be slower). A QT-temper treatment is often applied for strength, and a surface treatment to enhance corrosion resistance. Cold heading is strictly high volume, generating parts by the hundreds of thousands.

Because it is done cold, strain hardening plays a major role in this method, with peculiar consequences for form freedom. For instance, making a bolt with an internal hexagon socket requires strongly deforming an already-deformed part of the head. Consequently, this is only feasible with metals that have a low strain hardening exponent $n$, such as low carbon steels (to ensure sufficient hardenability, some extra alloying elements will be required, especially for larger fasteners). With stainless steel, for example, this would be prohibitively expensive because of excessive tool wear. In fact, the ordinary AISI 304 is not suited to cold heading at all, and in practice, stainless steel fasteners are therefore made of AISI 304 Cu, a variant of the same steel type containing 4% copper—an illustration of how a process choice influences the precise material choice.

## Coining

If cold heading of fasteners represents one extreme of the simple upsetting of a cylindrical billet, coining represents the other end: not long and thin but very short and squat. Like the previous method, coining is always done cold. It is instructive to study the process in some detail. Figure 6.11 shows how a disk-shaped "blank" is formed between two dies and inside a ring. So coining requires a set of three product-specific die elements. The blank itself is made by first punching it out of sheet metal, then rolling its edge so that this is slightly higher than the disk itself, and finally coating it: for instance, the common 50 euro cent piece is made of steel and coated with brass.

Two things of particular interest stand out here. Note that this is a true "closed-die" process, without flash formation, as the amount of metal in the blank is exactly that needed. The constraint and friction generate particularly high pressures in the die. The influence of friction is so large in fact that coining dies are slightly convex, so that contact and deformation starts at the center and works radially outward. If we used flat dies, the build-up of pressure toward the center would be so large that no deformation would occur there at all (recall Figure 6.6). Second, notice that coins are designed with positive, not negative, imprints *on* the coin. A negative imprint would require a positive pattern on the die, which would rapidly wear off. So the combination of high pressures with negative patterns on the die is

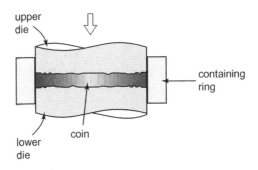

**FIGURE 6.11**

Schematic of the coining process. (For color version of this figure, the reader is referred to the online version of this chapter.)

essential for giving very high reproducibility, and the take-up of every fine feature of the pattern and lettering on the die faces. So even a seemingly simple "forging" such as a coin hides a world of refinement—and that is without going into the staggering production volumes. For the Netherlands alone, with some 16 million inhabitants, introducing the euro required an astounding 8 *billion* coins to be held in reserve (roughly 1 million coins per hour, if you wanted to produce them in one year).

---

**EXERCISE 6.18**

Take a close look at some coins. Why is there no parting line in the middle? Do you think the ring around the edge can be made out of one piece?

---

## 6.7 **Further Reading**

This chapter has explained the basics of forging, along with the main methods. The focus has been on processes that produce parts instead of semi-finished products (in which *rolling* of sheets, *rotary swaging* of tubes, *drawing* of wire, and *roll-forging* of profiles would become relevant additions). Furthermore, coverage has been brief of the various metals that are commonly forged; lack of mention here (e.g., for magnesium) does not imply that a metal is not forged—it just means that you are unlikely to encounter it as a forging in the normal line of your career. Recall too that forgings, like castings, are nearly always machined afterwards to bring them to final shape and tolerances. In practice, the industries of forging and machining are tightly linked. This sets the stage for potentially difficult trade-offs between forgability and machinability when it comes to choosing the right metals and alloys—especially when cost is factored in. It is hard to give clear examples of these issues that generalize well across different metals and applications, so this is one area where experience will have to be your guide. Consult the following literature if you would like to learn more details or discover more methods and forging metals.

Forging (and heat treatment) are covered in a number of journals with a wider remit, such as *Metal Science and Heat Treatment of Metals*, *Journal of Metals* and *Advanced Materials and Processes*, and the online industry periodical http://www.metalformingmagazine.com. For professional trade fairs, try the International Cold Forging Group (see http://www.icfg.info) and the United States–based Forging Industry Association (see http://www.forging.org)—this latter site includes a (fairly biased) "design engineering center," worth a look for interesting facts and figures.

Ashby, M.F., Shercliff, H.R., Cebon, D., 2013. Materials: Engineering, Science, Processing and Design, third ed. Butterworth-Heinemann, Oxford, UK (Chapters 19 and GL2). (*An introductory text at a complementary level to this book, providing detail on material properties and processing in a design-led context. Chapter 19 discusses friction and aspect ratio limits for forging; "Guided Learning Unit 2" includes a condensed treatment of the Fe-C phase diagram and phase transformations in steels, and hardenability.*)

Semiatin, S.L. (Ed.), 2005. ASM Handbook, Volume 14A: Metalworking: Bulk Forming. In: ASM International, Materials Park, Ohio, United States. (*Highly useful for professionals, this handbook remains the top reference work for forging processes, metals, applications, and process modeling.*)

Llewellyn, D., Hudd, R., 1998. Steels: Metallurgy and Applications, third ed, Butterworth-Heinemann, Oxford, UK. (*An excellent work for those wanting to know more about the world's most important structural material, with ample attention to heat treatments.*)

# Machining

$7$

## CHAPTER OUTLINE

Integrally machined from a slab of 7000-series aluminum, this high-performance brake for racing bikes showcases the design freedom of machining. Winner of the 2006 European Aluminium Award, this "BramBrake" outperforms its competition on stiffness and weight. Image courtesy M5 Ligfietsen.

## 7.1 **INTRODUCTION**

In machining, we start with a solid block of material and bit by bit remove material where we do not need it until we reach the desired shape or finish. There are four reasons why this manufacturing principle is of great relevance to design and engineering. First, it can be used to create finished components starting from a block, slab, or rod with standardized dimensions. Such semi-finished products can be obtained from stock or at short notice, requiring no part-specific investments, and have very predictable, homogeneous properties. Thanks to the advent of fast and fully automated machining centers, this "integral machining" is seen more and more, with various recent Apple products as exemplary applications. Second, machining is relevant because it can finalize the shape of a part made earlier by some other process, for instance, when we drill accurate holes into a casting or forging. In other words, machining is then used as a secondary shaping process. Another key role of the process is to make the complex molds and dies that in turn are used in many other processes. Finally, machining is also used to deliver a required surface finish.

As a manufacturing principle, machining is extremely flexible. The form freedom is virtually unlimited, and nearly all materials can be machined: metals, plastics and wood, and even glass and ceramics. Next, machining can produce single prototypes or small series (using universal equipment), but also large and very large volumes (using part-specific investments), and everything in between. Then there is the varied character of machined products. At first sight, the principle lends a clean, high-tech look and feel to products—though traditional sculpting and carving with hammer and chisel can also be seen as a form of machining, and one that is clearly artistic. Most machining methods are "mechanical"—drilling, turning, milling, grinding, and polishing—but machining also uses heat (including lasers), as well as water and chemicals! The range of material removal methods is wide, each offering certain geometric characteristics, precision, and finish.

Not surprisingly, function, cost, and quality are dependent on each other for machining, just as they are for other manufacturing principles. This chapter tackles machining from the perspective of the manufacturing process triangle and the underlying materials science, and explores what that means for design. We limit the discussion to *mechanical machining*, but having seen what to consider, the reader can be expected to approach other methods in a well-informed way.

---

**EXERCISE 7.1**

Look around your house and find one or more examples of machined parts or products. Try to find examples of integral machining, and machining as a secondary process.

---

## 7.2 **THE PRINCIPLE OF MACHINING: PROCESS BASICS**

Figure 7.1 shows the core of what mechanical machining is about. The leading edge of a cutting tool moves with constant speed through a solid slab of material, at a fixed depth below the surface. The cutting edge is sharp and triangular, with a certain internal "contour angle" $\beta$ and with its position relative to the material defined by the cutting edge "rake angle" $\gamma$, and the "relief angle" $\alpha$. The sum of

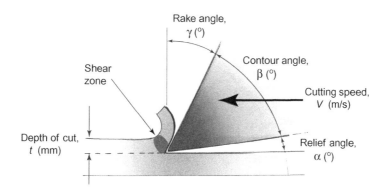

**FIGURE 7.1**

The idealized machining process. (For color version of this figure, the reader is referred to the online version of this chapter.)

these three angles as defined is exactly 90 degrees, and we can therefore freely select any two of them. (The contour angle can be larger than 90 degrees, in which case $\gamma$ becomes negative.) The topmost layer of the material, with cutting depth $t$, is removed, curling upward to form the machining swarf or "chips". This initial simplified cross-sectional view is two-dimensional, with no out-of-plane motion. In practice, inclined and curved tool edges and faces may lead to a more three-dimensional deformation pattern. We also assume for now that the cutting tool is sufficiently sharp and does not deform or break.

You would perhaps expect that machining is inherently a fracture process, with the chip being pulled off or split open, as when chopping firewood. Surprisingly, though, normal machining is a process of plastic deformation, localized into a small "shear zone" just ahead of the cutting tool, in which the material is sheared away by the tool (see Figure 7.1). The plastic mechanism of yielding in the shear zone (by dislocation motion) is exactly the same as takes place in a standard tensile test—though at a much higher rate of deformation—and the material quickly gets hot as it is removed. So for a first model we can expect that the higher the yield strength of the material, the more force this process will require. For isotropic materials, the shear yield strength is typically 50% to 60% of the yield strength. In machining ductile materials, we form long, continuous chips that curl up into spirals; with more brittle materials, we get short chips instead. So the strain-to-failure is also a factor in machining. Table 7.1 lists some relevant material data for various common materials.

---

**EXERCISE 7.2**

Sketch a small cube of material loaded by a shear force $F$ in N. How is the shear stress $\tau$ in N/m$^2$ defined?

---

**EXERCISE 7.3**

Why can we assume that for isotropic materials, shear strength rises with yield strength? And why not if the material is anisotropic?

**Table 7.1** Main Mechanical Properties and Cost of Various Materials

| Material | Yield Stress [MPa] | Young's Modulus [GPa] | Strain-to-Failure [%] | Approximate Cost (euro/kg) |
|---|---|---|---|---|
| Low carbon steels | 250-395 | 200-215 | 26-47 | 0.47 |
| Tool steels (AISI D2) | 1,860-2,290 | 205-215 | 14-15 | 4.50 |
| Aluminum alloy (6082-T6) | 240-290 | 70-74 | 5-11 | 2.00 |
| Brass (cast Cu-30Zn) | 90-120 | 93-95 | 23-50 | 5.45 |
| Magnesium (cast Mg-9AlZn) | 125-165 | 45-47 | 5-7 | 3.15 |
| Polyethylene (HD-PE) | 26-31 | 1.07-1.09 | 1,100-1,300 | 1.47 |
| Polycarbonate (PC) | 59-65 | 2.3-2.4 | 110-120 | 3.45 |

Source: Cambridge Engineering Selector (CES), Granta Design, Cambridge, UK.

**EXERCISE 7.4**

Judging from Table 7.1, which of the listed materials is easiest to machine, and which is the most difficult? Explain why.

**EXERCISE 7.5**

How does the strain-to-failure affect the machining process, apart from the chip length? Hint: remember the definition of the total plastic deformation energy in tension (in $J/m^3$) (see also Chapter 4).

**EXERCISE 7.6**

What do you think happens to the machining force that drives the cutting tool forward if we double the cutting depth?

Before digging deeper, we briefly consider two other materials: wood and stone. Most types of wood can be machined well, and the forces that are required are relatively small. However, parallel to the grain, wood is much stronger (and stiffer) than perpendicular to the grains, so wood is highly anisotropic. It can in fact split easily along the grains, so the failure mechanism under shear is different for wood than it is for metals or plastics. Stone also behaves differently; it is so brittle that it cracks easily. Therefore, stone can only be machined slowly, with low speeds and small cutting depths. If you take a closer look at a set of drill bits, you will notice that bits for metals, wood, and stone are quite different!

Sheet metal manufacturing processes such as blanking and punching also operate by exceeding the material's shear strength. Still, there is a key difference with machining, apart from the shape of the ingoing material. What is that?

## 7.3 DIGGING DEEPER: SPRINGBACK, HEAT AND LUBRICATION

Figure 7.2 takes a closer look at the machining process. First, notice how the cutting tool not only needs to be propelled in the horizontal direction relative to the workpiece (tangential force component: $F_h$), but that it also must be pushed downwards (normal force component: $F_v$). This normal component prevents the tool from "climbing" out of the material. The smaller the cutting edge angle $\gamma$ gets, the larger this component must be. Also, notice how right under the tool, the material deforms elastically under the normal force, and immediately behind the tool it springs back up. This is why we need to have the relief angle $\alpha$; without it, the friction between tool and material would be even greater. Plasticity theory tells us that, for typical cutting edge rake angles, the tangential force $F_h \approx 1.5\, k\, t$ (per unit width of cut), where $k$ is the shear yield stress (typically 50% to 60% of the yield stress), and $t$ is the cutting depth, as before. We can typically assume that $F_h \approx 3F_v$. Figure 7.2 shows the resultant machining force, equal in magnitude to $F_m = \sqrt{(F_h^2 + F_v^2)}$.

Which material from Table 7.1 will show the largest springback during machining? Hint: first find the machining force components, and then apply Hooke's law.

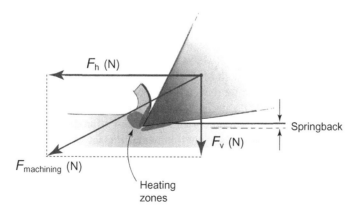

**FIGURE 7.2**

Forces, springback and heat development in the machining process. (For color version of this figure, the reader is referred to the online version of this chapter.)

---

**EXERCISE 7.9**

Briefly revisit Exercise 7.4. What is your answer now?

---

**EXERCISE 7.10**

Do you think that an alloy that is well-suited to sheet metal forming will also be well-suited to machining? Why or why not?

---

In practice, the relief angle $\alpha$ is around 4 degrees for steel, 6 degrees for aluminum, and 10 degrees for most plastics. Likewise, the optimal contour angle varies with the material. This implies that machining of material combinations can lead to uneasy compromises, for instance, when we have inserted steel bushings into an aluminum casting, that then have to be machined flat as one piece. During the design process, such combinations must therefore be viewed with caution.

Machining generates heat. This happens primarily inside the shear zone, where the material rapidly changes direction by plastic flow, but also by friction between the tool and material (i.e., on the tool face and, to a lesser extent, on the underside of the tool; see Figure 7.2). This heat is distributed through conduction and heats up both the material and the tool. In machining of metals, around 80% of all the heat ends up in the chips. This is advantageous, because we can then get rid of most of the heat together with the chips.

---

**EXERCISE 7.11**

Assume we use a milling machine (see also Section 7.6) with a 12 kW capacity to make a steel mold, removing 4 cm$^3$/s. If the specific heat of steel equals 470 J/(kg.K), how much hotter will the chips then get? Hint: you will also need the density of steel.

---

This previous exercise illustrates why some kind of cooling fluid is often used during the process, especially for integral machining (which involves rapid removal of relatively large amounts of material). The added benefit is that the fluid also washes away the chips, which can then be collected with filter systems and made available for recycling as production scrap. Fluids also act as lubricants, reducing the friction between tool and material. However, the chemicals used in machining lubricants can create an unpleasant working environment and may present an environmental disposal hazard. Machining often presents trade-offs in their environmental impact—lubricants enable much higher processing rates (reducing energy consumption) but at the penalty of a negative impact of chemical disposal.

Another development with the same goal of accelerating machining is the use of special alloy variants that enable higher speed machining. For example, "free machining steels" contain elements such as lead and sulfur that enhance machinability by promoting the formation of *inclusions*. These are usually somewhat detrimental to the resulting product properties, particularly toughness and fatigue. But in machining, the inclusions promote weakness in the chip shear zone, reducing chip size and cutting forces. The alloy constituents may also transfer to the tool cutting edge, acting as a tool lubricant. Lead is effective but is expensive and environmentally undesirable (or even banned—see Section 8.7). But again it can be argued that if cutting speeds are increased, and less damaging lubricant is used, there are environmental benefits—a typically difficult trade-off to quantify.

In practice, industry always aims to maximize speed. So the product of cutting depth, chip width, and speed must be maximal (i.e., in $m^3/s$). To be able to do this, the machine must first have sufficient power, equal to force times speed (i.e., $N \cdot m/s = J/s = W$). Furthermore, the cutting tool should not fail under the machining force. So the contour angle $\beta$ must be sufficient. However, keeping $\alpha$ constant, we know that the rake angle $\gamma$ decreases when $\beta$ increases. This increases the normal force, and with it, the machining force. So increasing $\beta$ is actually not effective. Reducing the relief angle $\alpha$ is also not an option: keeping $\gamma$ constant, this would increase $\beta$ but would also cause more friction under the tool and heat up both tool and material even more. All of this explains the need for extremely strong tool materials.

---

**EXERCISE 7.12**

Apart from its contour angle and strength, which other tool properties can you think of that are relevant to machining?

---

Large, powerful machines have been around for decades now, and suitable tool materials are also widely available. Still, machining cannot be done at just any speed: the product itself must also be able to deal with the machining forces. This is explained in the next section.

---

## 7.4 DEFORMATION OF WORKPIECES AND MACHINING TOOLS

Through the years, machining has increased steadily in terms of the accuracy that it can achieve. Micrometer precision, once the stuff of dreams, is now routine. However, certain geometries are still much easier to manufacture than others, in terms of achieving a high speed with reproducible accuracy and finish. This can be understood, at least in a qualitative manner, by considering how the part and the machine tool deform under the machining force, including the influence of the clamping arrangements.

First, let's consider part deformations. As an example, Figure 7.3a shows how a square, slender workpiece with length $L$, width $W$ and height $H$ deforms under the machining force $F_m$ if we do not support it—a highly undesirable and therefore unlikely situation, as we shall soon discover. We can decompose $F_m$ into two components: $F_h$ along the length of the workpiece, which compresses it a very small distance $u$ (not shown in the figure), and $F_v$ downwards, which bends it a much larger deflection $\delta$. Both $u$ and $\delta$ are elastic deformations, so when the force is removed, the material springs back, and the machined part will therefore deviate from the target geometry. For $L \gg W$ and $H$, the maximum bending deflection is given by $\delta = F_v L^3/(3EI)$, in which $I = 1/12\ W H^3$ for a section of width $W$ and depth $H$ (with $H$ being the dimension in the direction of the transverse load).

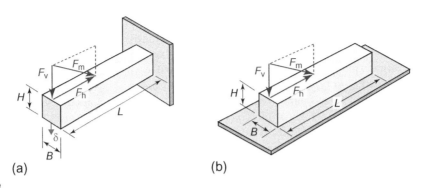

**FIGURE 7.3**

Machining (a) without, and (b) with, full workpiece support. (For color version of this figure, the reader is referred to the online version of this chapter.)

**EXERCISE 7.13**

A workpiece made of aluminum 6082 (as in Table 7.1) has dimensions $L=300$ mm and $W=H=30$ mm, and is machined as in Figure 7.3a. What is the downward force component $F_v$ needed to cause a deflection $\delta=0.1$ mm? What then is the typical accompanying component, $F_h$? Given a tool width of 5 mm, what cutting depth can we then use? Will this be productive, do you think?

**EXERCISE 7.14**

Now suppose that we have the same cutting geometry, but with the workpiece fully supported, as in Figure 7.3b. If we assume that the tool applies the same tool forces used in the previous exercise, explain why the deflection of the surface will be orders of magnitude smaller in this case.

We see that bending, especially in slender workpieces, can cause significant elastic deformations even at modest machining forces (torsion performs even worse in this respect). Conversely, pure tension and compression cause much smaller deformations and therefore allow much higher machining forces for the same tolerances, for much higher productivity. Now, in simple situations, we could perhaps anticipate such workpiece deformation, and somehow compensate for it by careful manipulation of the cutting tool. This is, however, difficult, even with expensive equipment. A better option is to ensure that workpiece bending (and torsion) under the machining force is avoided or minimized. If standard facilities for support and clamping cannot achieve this outcome, then you will need to design specific solutions. Apple's designers reportedly spent a significant amount of time developing the clamping systems needed for the manufacturing of the MacBook's aluminum "unibody" with its many thin details—premium performance, but at a premium price! But if you design your part well, you can generally avoid, or at least minimize, such problems.

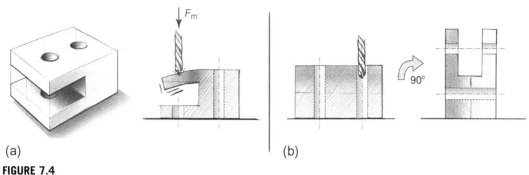

(a)                                                                              (b)

**FIGURE 7.4**

A component requiring two machining operations, showing (a) the wrong sequence, and (b) the correct sequence. First drilling both holes and then removing the inner section avoids bending. (For color version of this figure, the reader is referred to the online version of this chapter.)

The principle of using pure compression of the part as a reaction against the machining force, while avoiding bending, can also determine the sequence in which different machining operations should be executed. Figure 7.4 illustrates this via an example.

---

**EXERCISE 7.15**

Suppose we want to machine a thin-walled box-like shape out of a solid block of material (much like the iBook unibody). What do you recommend: first machining the walls to their approximate thickness and then finishing them, or finishing them directly?

---

Let us now briefly consider tool deformations. It will now be obvious that machining tools (milling heads, chisels, drills, etc.) deform under the forces they deliver as well, and furthermore, that slender tools under bending will deflect much more than tools under compression. Thick, squat tools suffer much less from this but may pose geometric restrictions of their own: for instance, a round milling head with a certain diameter cannot make internal radii smaller than that diameter.

---

**EXERCISE 7.16**

Dimensional inaccuracies can be caused by several other factors than the elastic deformations of tool and part. Name at least two. What can we do to minimize their effect? And are these measures related to design or to manufacturing?

---

Industrial manufacturing practice has adopted an additional solution to the problem of deformation: *high-speed machining*. The basic idea is simple. Instead of using a certain large force in combination with a low cutting speed (i.e., slowly machining with a comparatively large tool width and cutting depth), the force is much reduced by choosing a smaller tool width and cutting depth. Productivity is maintained simply by increasing the speed of cutting. One key to enabling this solution is the advent of ceramic cutting tools that can withstand the accompanying higher temperatures with ease; improved use of cooling fluids has helped as well. The brittle nature of these new tools demands extra care in the initial positioning, as simply slamming the tool against the workpiece could cause it to break. Computer control over the tool position is therefore another piece of the puzzle.

Does this now eliminate all restrictions on design? No, unfortunately not: function, quality, and cost are still closely linked. In high-speed machining, one particular new problem is that of resonance (i.e., the tool and/or the workpiece vibrate excessively, leading to loss of accuracy). This is a particular issue whenever thin-walled or slender parts are machined, or when delicate, thin tools are deployed to machine away the complex inside of a part, and so on.

## 7.5 ROUGHNESS AND SURFACE DEFECTS OF MACHINED PRODUCTS

Apart from tolerances, another quality attribute that applies strongly to machining is surface roughness. To understand how, we must expand Figure 7.1 into the third dimension and see what takes place there. In machining, the depth and width of cutting are always small relative to the part dimensions. So to remove a continuous surface layer, the various cuts, or grooves, must overlap smoothly with one another. But this never happens perfectly—for instance, because of deformations of workpiece and tool—and we end up with a repeating surface pattern. A common measure for this roughness is some average of the statistical fluctuations in height between the highest and the lowest points of this pattern (such as the "root-mean-square" height difference from the mean surface position). This $R_a$ *value* is typically given in micrometers and was traditionally measured with a fine-tipped "stylus" tracking over the surface irregularities, but today this result is routinely achieved automatically over large areas using laser interferometry.

If the roughness is too large for the desired application, then we can reduce the depth and width of cutting—but this will, of course, reduce productivity. A solution is to first machine the product using a big tool to quickly reach its approximate dimensions and then exchange the tool with a smaller one to finalize the product. But if the new tool is not accurately positioned, then we lose geometrical tolerance. A compromise approach is to use one tool but to reduce the size of the final cuts to give the required finish.

---

**EXERCISE 7.17**

How exactly does the strategy of first using large tools and high machining forces, then smaller tools and low machining forces, affect the geometrical tolerances? What is the consequence in terms of cost?

---

When dealing with high-speed machining, there is another potential problem: any vibrations that can occur when producing thin sections or part details not only affect part accuracy but can also cause undue roughness. This may inform the order in which sections are finished.

---

**EXERCISE 7.18**

For a high-speed machined, thin-walled metal electronics housing, what do you recommend: first finishing the inside, then the outside—or the other way around? Why?

---

Machining can also create certain characteristic surface defects, such as "burrs." At some point, the cutting tool will "run out" across an edge of the component and in doing so, it often leaves a sharp burr of material on the surface. Solutions to de-burr quickly and cheaply are therefore valuable, but it is even better if the defect is in a location where it can do no harm and can simply remain in place. Proper design for machining can achieve this goal. A second surface defect is a "lay pattern," a repetitive impression on the part surface, created by the tool (for an example, think of a freshly mown soccer field, on which you can usually see how the mower moved). On milled parts particularly, this is often clearly visible. Minimizing the lay requires either better-quality (stiffer) machining equipment and tools, which increases costs, or reducing cutting depth (and hence productivity). It is better if the pattern is not avoided, but instead used deliberately—the next section provides an example of how this can be done.

Finally, note that the roughness and absence of surface defects that can normally be achieved are not independent of the product size. For instance, the metal unibody of a modern cell phone will be, as it needs to be, much smoother than a truck engine block. Machining handbooks generally contain tables listing the attainable roughness and tolerances against product size. One guiding principle to remember in design, with regard to both tolerance and roughness, is that it is always costly to over-specify these quantities—in terms of both target values and the proportion of the component surface to which they apply.

---

**EXERCISE 7.19**

Take a close look at your product samples (see Exercise 7.1). Which quality issues can you spot, in terms of roughness, burrs, or lay?

---

## 7.6 MACHINING METHODS

Methods for machining are traditionally distinguished on the basis of the tool motion relative to the part. For design purposes, however, it is more useful to sort methods on the basis of the shapes that they can realize. *Turning*, for instance, produces rotationally symmetrical parts, *drilling* limits us to putting holes into existing parts and products (i.e., it is exclusively a secondary forming process), and *milling* allows us to make all kinds of shapes, depending on the number of axes along which tool motion is controlled. This trio is presented next. So-called *machining centers*, which combine these three processes, are distinguished here as a fourth method. Note that in all methods, the chips can in principle be collected and recycled. In large-scale machining operations, where the necessary specific investments can be recouped through the scrap material value, this is, in fact virtually always done (for metal, wood, and, increasingly, for plastics). Economically and ecologically, this simply makes good sense. However, scrap recovery is never 100%, so a certain fraction of the material is always lost. Apart from that, reprocessing scrap also takes energy (though always a fraction of the energy "embodied" in virgin material), and there is always the issue of scrap contamination (e.g., mixing of different alloys and entrapment of dirt and impurities) to contend with. Chapter 14 discusses these issues further.

## Turning

Many parts of common products, such as handles or axles, have rotational symmetry. For such parts, turning is the ideal machining method. Internal or external screw threads can be added, as well as visually appealing criss-cross "knurling patterns" for better grip (seen, for example, on metal hand tools and flashlight housings). These are, in fact, a good example of a lay pattern being applied deliberately as a surface *effect*. Figure 7.5 shows the archetypical turning machine: the *lathe*. The ingoing material—a rod, usually—is clamped on one side in the chuck and is fixed on the other end with a center pin. The lathe turns it at high speed (typically 1,000 to 5,000 rpm; the larger the chuck, the lower the speed). The circumferential cutting speed is determined by the rotation rate and part radius, whereas the removal rate is determined by the speed at which the tool is moved, either parallel to the rotation axis or radially toward that axis. Lathes come in all kinds and sizes, from tabletop to very large, and can be completely manually controlled, have full computer numerical control (CNC), or anything in between.

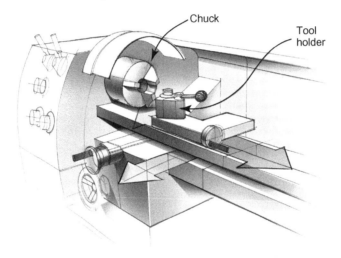

**FIGURE 7.5**

Lathe (classical model, without CNC operation).

---

**EXERCISE 7.20**

Even when clamped in the chuck and centered on the other end, the workpiece in a lathe will still bend under the machining force, if the cutting tool is operating somewhere in between—so we get comparatively large deflections for small forces. How can we solve this problem?

## Drilling

From the small handheld power drill you may have at home to huge drill presses in industry, drilling is still the best way to make a round hole in a product. An interesting fact about drilling is that the cutting edge angle $\gamma$ varies over the drill radius, from very small near the tip to much bigger near the outer

edges. So most of the force needed to push the drill down is due to the drill tip, and once that tip protrudes through the material, the drill tends to shoot right through. In metal, this often leads to a nasty
burr; in wood, it may break off splinters (which is why wood drills have long, sharp tips). In "blind
holes," which do not go all the way through the material, this problem is, of course, non-existent—
and skilled woodworkers know that the way to avoid splintering is to drill through the workpiece into
a backing block.

Even just for metals, there are all kinds of drills. Four kinds that every designer should recognize are
the center drill, the standard drill, the countersink drill, and the reamer (the last one only being needed
for very accurate holes, i.e., for fittings). They are often used in this sequence for a single hole, with
the addition that large holes (e.g., > 20 mm for steel parts) must be made in several steps. For normal
holes in ordinary materials, standard drills can make holes up to eight times deeper than their width
(i.e., $L/D \leq 8$). So-called spade drills can make deeper holes but require special preparatory steps
and, hence, are less productive; conversely, "jobber drills" can make shallow holes ($L/D \leq 2$) faster
by dispensing with the separate use of a center drill.

---

**EXERCISE 7.21**

Look around in a local workshop or find detailed illustrations of these six kinds of drills on the Internet. What are the
key differences (e.g., how many cutting surfaces do they possess)?

---

## Milling

Again, the machine types found under this method are varied, but they all share a common layout: a flat
working table on which a block of material can be clamped, with a milling cutter above it holding the
actual cutting tool. The tool rotates at high speeds with considerable down force. By moving the table
relative to the cutter, on multiple axes, we can remove material from the block and produce all manner
of shapes. The more degrees of freedom (DOF) the milling machine has, the more form freedom it
offers without having to reposition and reclamp the product. That is positive, because reclamping takes
time and reduces accuracy: we then get the precision of clamping, not the precision of the machine
itself, which is much higher (alternatively, we can readjust the machine carefully, but this takes even
more time). The downside of machines with many DOFs is that they are more expensive to operate and
less flexible with respect to product size.

A milling machine always has at least two axes of motion, and usually at least three: the workpiece
can always move up and down (this is referred to as the $z$-axis), and laterally from side to side ($x$-axis),
plus in some cases from front to back ($y$-axis), giving a triaxial milling machine. If we add the ability to
rotate the cutter around the $x$-axis, we get a four-axes machine; rotation around the $y$-axis gives us a
fifth DOF (Figure 7.6). Milling can be done as "climbing" or "conventional": in the first case, the rotation of cutter and feed are in the same direction at the location where material is removed; in the second
case, the motions are in opposite directions. Generally, climbing milling is preferable for hard materials, conventional milling for soft ones.

Just as with drilling, milling involves several kinds of cutting tools. The main three are the end mill,
the slab mill, and the ball nose mill:

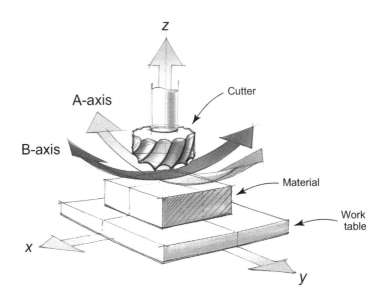

**FIGURE 7.6**

Degrees of freedom (DOF) in milling. NB: rotation around the x-axis is the A-axis DOF; rotation around the y-axis is the B-axis DOF. (For color version of this figure, the reader is referred to the online version of this chapter.)

- The end mill is shaped like a long, narrow cylinder, typically with two or three cutting edges. Its main purpose is to create fine details and narrow slots. Often such tools also have cutting edges on the underside, like a drill, so that they can cut vertically into a block, then move horizontally to create the desired shape, with no need to pre-drill a hole.
- The slab mill is also cylindrical but now short and squat, typically with five, seven, or nine cutting edges. Its main strength is its capacity to remove large amounts of material fast. Such tools have no cutting edges on the underside. Note, however, that a machine with at least four axes can push a slab mill into the material at an angle, also eliminating the need to pre-drill (called "ramp-in"). Of course, this requires CNC operation.
- The ball nose mill has, as its name implies, a spherical head. It main purpose is the manufacture of double-curved shapes. Note that its capacity, in terms of $cm^3/sec$ it can remove, is much lower than that of the previous two kinds, so for single-curved shapes its use is to be avoided.

---

**EXERCISE 7.22**

Suppose we use a triaxial milling machine equipped with an end mill (with a 10 mm diameter) to hollow out a rectangular block. What is the minimum internal radius of the hollowed-out block?

---

**EXERCISE 7.23**

Think of a shape that can be made on a CNC-operated milling machine with four or five axes, using a standard end mill only, that *cannot* be made on a triaxial machine, without re-clamping the product (at least not with acceptable productivity).

> **EXERCISE 7.24**
>
> Why is a ball nose mill needed for double-curved shapes? And why do such shapes require CNC operation over at least three axes? Explain.

## Machining centers

A *machining center* combines the three methods discussed earlier. Such machines can switch tools between operations automatically as needed and can therefore work from coarse to fine, combining speed with accuracy. Some can even reclamp the product as desired, and all of that with *computer-aided design/manufacturing* (CAD/CAM) linkage. They can be sufficiently productive even for large production volumes. Still, even these "black box machines" obey the physical laws described in this chapter. Proper design therefore remains the best guarantee for a good result.

Total machining time is often decisive for the costs of this method. For normal-sized machines operated in two-shift labor, machine costs are around 60 to 65 euro/hour. For a part such as the Apple MacBook's aluminum unibody from 2008 and later, which requires an estimated production time of 20 minutes (considerably longer than casting would take!), these costs can be considerable and easily higher than the material costs. Manufacturing setup costs, including those for any dedicated clamping and fixing systems, add to the eventual part price, but as a general design rule we can tentatively state that the total machining time must be minimized.

> **EXERCISE 7.25**
>
> Look back at one of the product examples you found (see Exercise 7.1). Consider which metal and method were involved, and then draw up three design guidelines for this combination. How do you see these rules reflected in your product?

## 7.7 Further Reading

This chapter only covers mechanical machining, omitting less common methods, such as *planing* and *boring*. Non-mechanical processes exist also: examples are electro-erosive machining and chemical etch machining. However, the manufacturing triangle and many of the main concepts outlined for machining in this chapter equip you to research and understand these processes too. Consult the following sources listed—and do not forget to take a look in an industrial machining workshop!

Innovation in mechanical machining marches on. Industry keeps developing new alloys that combine hardness and ductility with good machinability, such as the *Toolox* steels from Swedish Steel AB. Such material developments are not just economically driven: for instance, the recent EU directive *RoHS* (from *Reduction on Hazardous Substances*) now forces industry to find lead-free alternatives for conventional lead-based machining alloys of steel, brass, and aluminum. Here it is not just finding new materials that can be costly and time-consuming, but especially *proving* that they work as well as the materials they replace (i.e., re-certification of parts).

Meanwhile, the accuracy of machining processes keeps getting better and better. Already, commercially available machines exist that can routinely reach a surface roughness $R_a$ of just 0.2 μm in heat-treated, double-curved steel products (e.g., dies for metal casting, molds for plastic injection molding) at impressive speeds. In grinding and polishing processes, which are basically machining processes at the microlevel, this development is equally relentless and is getting astoundingly close to the atomic level—as predicted by the Japanese machining expert Norio Taniguchi back in 1974.

The International High Speed Machining Conference, held annually, is one of the leading industry events for the exchange of information in the field of machining (see www.highspeedmachining.org). Scientific journals on the topic include *Machining Science and Technology* and the *International Journal of Machining and Machinability of Materials,* though neither directly focuses on design.

Davis, J.R. et al., (Ed.), 2002. American Society of Metals Handbook, Volume 16: Machining. ASM International, Materials Park, Ohio, United States. (*A comprehensive, world-leading reference volume on the subject. Expensive, but worth the investment if you are serious about machining.*)

Oberg, E., et al., 1996. Machinery's Handbook, twentyfifth ed. Industrial Press Inc, New York, United States. (*Hard to get, but still "the bible" for true machining experts.*)

# Injection Molding of Thermoplastics

Obtaining an honorable mention in the 2013 Red Dot Design Competition, the Dopper is an attractive and functional re-usable water bottle, providing a sensible and stylish alternative to disposables. Its cap and cup are injection molded, while the bottle itself is first injection molded as a pre-form ('parison') and then blow molded.

*Image courtesy Dopper.*

## 8.1 INTRODUCTION

Without any doubt, injection molding is *the* most important manufacturing process for the field of industrial design. The main reasons are its high production speed, its huge form freedom, and its unparalleled potential for functional integration. Also, injection molding is suitable for a class of materials that is a world of variety in itself: that of thermoplastic polymers, or "plastics" for short. These materials not only span a broad range of mechanical properties, but they can also have all kinds of colors and levels of transparency. With suitable mold design, even their texture can be varied widely. And then there are "grades" of plastic that release a special smell, shield electromagnetic fields, or feel surprisingly cold to the touch.

As usual, the design freedom comes at a price. Injection molding is not possible without product-specific molds, and these tend to be relatively expensive. Just how expensive they are depends, among other things, on the product size and complexity and on the desired accuracy. We must note that many developing countries (not just China) have embraced both molding and mold making enthusiastically, causing the price of seemingly identical molds to vary by a factor of five worldwide! But one way or another, this manufacturing principle shows a strong interdependence of function, cost, and quality. For instance, certain form details are much easier (that is, cheaper or more accurate) to realize than others, even with optimal mold design. If you try to capture this complex reality in a few easy design rules, you would not do any justice to the diversity: there are tens of different manufacturing methods for injection molding, each with unique possibilities and limitations. Therefore, it is best to study the principle in detail first. In this effort, we limit the discussion to injection molding of thermoplastics.

---

**EXERCISE 8.1**

Look around your house for injection molded parts and products, and begin to consider how important the following might be in their design: form freedom, color, surface texture, stiffness, strength, or cost.

---

## 8.2 THE BASIS: THERMOPLASTIC BEHAVIOR

Thermoplastics, you will recall, are *polymers*: very long chains of basic hydrocarbon building blocks linked together. At the micro-level of the material, these chains are entangled in a manner that can be compared to a portion of cooked spaghetti. It is this chain entanglement that forms the basis of the mechanical and physical properties. Polymer chains are attracted to one another mainly by Van der Waals forces, which are considerably weaker than the interatomic bonds in metals, and consequently, the base stiffness of polymers is low: for example, for the common plastic ABS, stiffness is typically just 1 to 2 GPa, compared to 70 GPa for aluminum. However, the entangled mass resists tearing and rupturing very effectively, as this requires the strong covalent bonds *along* the chains to be broken. The base strength and strain-to-failure are therefore surprisingly high: easily 40 MPa and 50% for ABS. This requires the polymer chains to have a certain minimum length, which can be quantified either in terms of the number of monomers in the chain or its total molecular weight. The building blocks themselves, called *monomers*, can have all manner of shapes, starting from a simple chain of carbon, each with two hydrogen atoms attached, up to complex carbon chains with hydrogen, chlorine, and side groups of other hydrocarbons. Indeed, the chain itself need not consist of carbon atoms only: for instance, in polyamides ("nylons") we find some nitrogen in it as well. The chain chemistry and shape strongly influence the material properties, including, for instance, whether a thermoplastic is *amorphous* or *semi-crystalline*, which is an important difference. Even more variety comes from the different groups of additives that can be applied to a thermoplastic material (henceforth simply called *plastics*), such as pigments for color, flame retardants, fillers for low price, glass fibers for strength and stiffness, or plasticizers for faster production.

---

**EXERCISE 8.2**

How many monomers typically constitute a polymer chain?

---

**EXERCISE 8.3**

Sketch the main differences in chain structures for a thermoplastic material, a thermosetting material, and a rubber. Explain these differences in a few keywords.

---

**EXERCISE 8.4**

There are six 'commodity plastics', each with its own recycling logo. Which ones are they, and what are their monomers? Name a typical application for each one. Are they semi-crystalline or amorphous, or can they be both, depending on the specific grade?

---

**EXERCISE 8.5**

Next, name five 'engineering plastics' (defined here as all non-commodity plastics with prices below 10 euro/kg). Give typical applications, and indicate if they are semi-crystalline or amorphous.

---

When we slowly heat up an amorphous plastic, then over a small temperature range we see something remarkable: the stiffness drops enormously (by a factor of order 100 to 1,000) and the material assumes a rubber-like consistency. The temperature at which this happens is known as the *glass (transition) temperature, $T_g$*. The explanation is that above this temperature, the molecular chains have sufficient freedom of motion to slide along each other, which was not possible at lower temperatures (the strength of the Van der Waals forces depends strongly on the distance between molecular chains). Figure 8.1a illustrates this behavior—note the logarithmic scale for stiffness. Above $T_g$, amorphous plastics cannot normally be used, but they now can be thermoformed (see Chapter 9). With increasing temperature, the stiffness continues to drop, gradually at first, then more rapidly around the *softening*

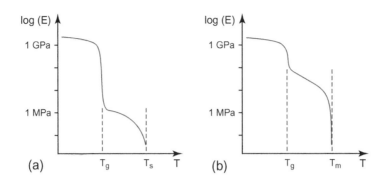

**FIGURE 8.1**

Stiffness versus temperature for (a) amorphous plastic and (b) semi-crystalline plastic. (For color version of this figure, the reader is referred to the online version of this chapter.)

*temperature*, $T_s$. Indeed, beyond $T_s$ the concept of stiffness no longer applies: the plastic becomes a *melt* and can now be injection molded.

In semi-crystalline plastics, the glass transition is less pronounced, because the crystalline areas only lose their strength at the *melting temperature*, $T_m$ (which is typically some 50% higher than the $T_g$, in Kelvin), keeping the material together until then (see Figure 8.1b). Thermoforming such materials is difficult, or even impossible in practical terms if the degree of crystallinity is very high (again, see Chapter 9), but beyond $T_m$ the material can be injection molded. Note that unlike amorphous plastics, semi-crystalline plastics are normally not used at temperatures below $T_g$, as they are then too brittle. Polypropylene (PP) is an interesting case to mention here: its $T_g$ is around 0°C, making pure PP unsuited for application in everyday objects, which is why it is normally copolymerized with 5% to 10% polyethylene (PE). The much lower $T_g$ of PE ensures that the material is sufficiently tough, even at sub-zero temperatures.

---

**EXERCISE 8.6**

What are the values of $T_g$ for the six commodity plastics? What are their maximum service temperatures?

---

For a polymer melt, viscosity values of the order of magnitude of 10 to 100 Pa.s (Pascal seconds) are common. If we compare this with, for example, molasses, which has a viscosity of around 5 Pa.s, or with the 0.001 Pa.s of water, we see that polymer melts are viscous indeed. This high viscosity, combined with the narrow sections that the process allows us to design, is the main reason why injection molding requires very high pressures, and hence why machines and molds are so expensive. Just how much pressure is needed is the subject for Section 8.3.

Exactly how viscous the melt will be depends mainly on its average molecular weight, $M_w$: the longer the chains, the thicker the melt. In practice, this translates to a trade-off between processability and, especially, toughness. Furthermore, viscosity also drops slightly with increasing temperature. Higher temperatures therefore allow for somewhat lower injection pressures. However, if we increase the temperature too much, the plastic will degrade thermally, causing ugly discoloring and reducing the mechanical properties. Also, the cooling part of the injection molding cycle takes more time if the melt temperature is higher. Typically, injection molding is done at around 100°C beyond $T_s$ (amorphous) or $T_m$ (semi-crystalline). In all of this, we, of course, already see some principal trade-offs in the manufacturing triangle between function, quality, and cost.

---

**EXERCISE 8.7**

Which kinds of plastics have been used in your sample products from Exercise 8.1? Tip: Ensure that your answers match what you said in Exercises 8.4 and 8.5!

---

**EXERCISE 8.8**

Thermosets and rubbers do not melt on heating. What happens instead, and why is this? What are the implications for the recyclability of these polymers compared to thermoplastics?

---

## 8.3 FILLING THE MOLD: PRESSURES AND CLAMPING FORCES

To understand the core principle of injection molding, imagine what happens if a melt flows through a flat, straight channel (Figure 8.2a, with the melt flowing from left to right). Two locations are shown, Point 1 and Point 2, with local pressures $p_1$ and $p_2$ (with $p_1 > p_2$), and positioned a length $l$ apart. We assume the channel has thickness $d$ and width $b$ ($\gg d$, so that there is no variation in the out-of-plane direction). Notice how we have a certain velocity profile over the thickness, from zero on both walls to a maximum in the middle. This means that the velocity gradient $dv/dy$ also varies, from maximum at the walls to zero in the middle: this gradient we call the *shear strain-rate* (units: $s^{-1}$).

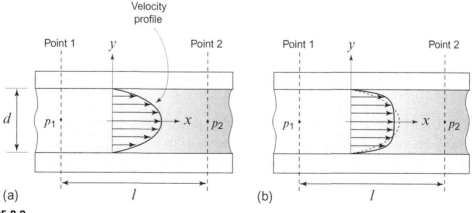

**FIGURE 8.2**

Velocity profile, with flow from left to right, for (a) Newtonian fluid and (b) non-Newtonian fluid (dotted line = profile from part a). (For color version of this figure, the reader is referred to the online version of this chapter.)

We now pose the question, how much pressure difference $\Delta p = p_1 - p_2$ do we need to apply to overcome the friction forces in the melt and push it all the way from Point 1 to Point 2? For normal fluids we can answer this question with relative ease because the relationship between the shear stress $\tau$ and the shear velocity is linear: $\tau = \mu \, dv/dy$, where $\mu$ is the viscosity (units: Pa.s). If this equation holds with a constant value of $\mu$, then we are dealing with a "Newtonian fluid," and the accompanying velocity profile is quadratic. By integrating the shear stress over the thickness, we get the pressure difference per unit of flow length. Multiplying by length $l$ we then get

$$\Delta p = \frac{12\mu}{t_{in}} \left(\frac{l}{d}\right)^2 \tag{8.1}$$

where $t_{in}$ is the injection time—that is, the time required for the melt to fill the length of the channel (the shorter this time, the more pressure we need). As usual, there are several assumptions behind this simple equation, apart from the fluid being Newtonian: we assume that the flow is laminar, not turbulent, and we ignore what happens at the beginning and at the end of the channel (valid for $l \gg d$). Regarding the former, we note that injection molding typically is done at flow speeds around 1 m/s and, combined with the high viscosity of the melt, this ensures laminar flow. Note that Equation (8.1) is only valid for the geometry and type of flow in Figure 8.2a, known as *Poiseuille flow*. Do not use it for other kinds of flow! For example, for the case of filling a disk-shaped cavity, enclosed by two plates and filled through a single, central, circular gate, the pressure difference is given by

$$\Delta p = \frac{6\mu}{t_{in}} \left(\frac{R_2}{d}\right)^2 \ln\left(\frac{R_2}{R_1}\right) \tag{8.2}$$

where $d$ is the disk thickness, $R_2$ the disk radius, and $R_1$ the radius of the gate; the other quantities are the same as in Equation (8.1), and again it is only applicable to Newtonian fluids. To give a simple numerical example for Equation (8.2), consider a disk with a thickness $d = 2$ mm and a radius $R_2 = 100$ mm. If we inject using a circular gate of radius $R_1 = 5$ mm, a plastic with viscosity $\mu = 36$ Pa.s, and an injection time $t_{in}$ of 0.2 s, then we need a pressure difference $\Delta p = 39$ MPa, or 390 times atmospheric pressure!

---

**EXERCISE 8.9**

Determine the pressure difference that is required to fill a cell phone housing in one tenth of a second. Model the shape as a flat plate that is injected over its full width, assume Newtonian behavior, and use a viscosity $\mu = 36$ Pa.s.

---

**EXERCISE 8.10**

What happens to the pressure difference if—at equal injection time—the thickness is halved?

---

So far, we have only considered the pressure difference that is needed to overcome the viscosity. Applying just this amount of pressure at location 1 would mean that the pressure at location 2 is zero. This is to be avoided, as the surface tension of the melt would then prevent the material from filling tight radii and accurately reproducing fine mold details, such as mold texturing, lettering, and logos. Squeezing the melt into tight corners and small details can easily require an additional 100 bars (10 MPa) or so. (Section 8.6 will show that there is a second reason why this extra pressure is needed.)

In our analysis we have assumed that polymer melts behave as Newtonian fluids (i.e., as having a viscosity that is independent of the shear strain-rate). This simplification may allow for comparatively easy modeling, but it is not reality: real polymer melts are decidedly non-Newtonian in behavior. Specifically, with increasing shear strain-rate, their viscosity drops significantly, as shown in Figure 8.3 for the commodity plastic polystyrene (PS). Observe the middle curve for 200°C: at a shear strain-rate of $10^1$ s$^{-1}$ the viscosity is some 1,000 Pa.s, whereas at a shear strain-rate of $10^4$ s$^{-1}$ it drops to around 10 Pa.s. This phenomenon is known as *shear thinning*. Note that we can now only speak of "the" viscosity for a certain shear strain-rate. This also explains why the velocity profile is different in Figure 8.2b: for the same average flow velocity, the peak velocity is lower in the non-Newtonian case.

During actual mold filling, shear strain-rates are of the order of $10^3$-$10^4$ s$^{-1}$ for most parts of the mold and during most of the injection. In this domain we may approximate the dependency of viscosity on shear strain-rate as linear in a double-logarithmic plot, leading to the so-called power law models—for details, consult the Further Reading section. The take-home message is that, unless you manage to choose your average viscosity really well (as we attempted in Exercise 8.9), any estimation of the pressure difference that assumes Newtonian behavior will likely be quite far off the mark, although it remains usable as a means to build up basic insight into the effects of the main parameters.

Returning to Equation (8.1), we can see that the thickness (or, more precisely, the ratio of $l/d$) has a very strong influence on the pressure differences required. This is the case for both Newtonian and non-Newtonian fluids. Also, the precise grade of the plastic is important: specifically, the longer the chains (i.e., the higher the average molecular weight), the higher the viscosity at a given shear strain-rate and temperature. Professional injection molders refer to this last property by means of the *melt flow index* (MFI), expressed in g/min. Put simply, this index is the inverse of viscosity, for very low shear strain-rates. (The fact that mold filling typically involves *high* shear strain-rates does not help the usability of the MFI, but it is an industry standard nonetheless.) The higher the MFI, the better the grade that can be processed. In practice, the MFI varies between 3 g/min for grades that are very hard to process to 30 g/min for "ultra-flow grades."

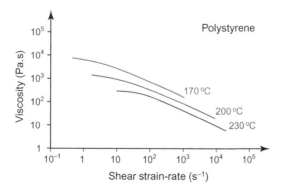

**FIGURE 8.3**

Viscosity versus shear strain-rate for polystyrene at different temperatures (log scales). (For color version of this figure, the reader is referred to the online version of this chapter.)

---

**EXERCISE 8.11**

Low molecular weight grades require less pressure to process than high-molecular-weight grades. But what disadvantages do they have?

---

Figure 8.3 shows that viscosity not only depends on the shear strain-rate but also on the temperature, as noted in the previous section. The higher the temperature, the lower the viscosity.

---

**EXERCISE 8.12**

Repeat Exercise 8.9, but now using a 10% lower viscosity thanks to increased processing temperature. What is the gain in terms of the required pressure difference? Would this increase also have drawbacks?

---

In injection molding, much as in high-pressure die casting of metals, we use two mold halves that are pushed together with a certain clamping force, with the cavity machined into the mating faces of the mold. This plane where the two halves meet is called the *parting plane*. The minimum clamping force (in N) is equal to the pressure difference (in $N/m^2$) times the effective surface area (in $m^2$) (i.e., the projection of the cavity onto the parting plane). (In practice, molders often use more clamping force to reduce any risk of flash formation, i.e., molten plastic creeping between the mold halves.) The higher the clamping force, the larger the injection molding machine we need, so it makes sense to be able to estimate this force.

---

**EXERCISE 8.13**

Calculate the minimum clamping force necessary for injection molding of the cell phone housing from Exercise 8.9. Tip: First sketch the position of the parting plane! Now do the same for one of your sample products (Exercise 8.1, repeating the estimate of Exercise 8.9 for this product).

---

**EXERCISE 8.14**

Suppose that to injection mold a round 10 litre PP bucket we require a pressure of 900 bar. How large should the minimum clamping force be?

---

As stated before, the equations given in this section only apply to flows through simple, straight channels. When dealing with complex products, accurate prediction of the required injection pressure and, hence, clamping force can only be done with computer assistance. And of course, we must take into account the effect of shear thinning. However, making a first estimation is certainly possible and comes in handy for gaining physical insights—and for checking that the elaborate computer code is working! Every designer should therefore be able to make such "back of the envelope" calculations.

## 8.4 A CLOSER LOOK AT THE INJECTION STAGE

In injection molding of plastics, just as in high pressure die casting of metals, the cavity is nearly always filled through a number of small, narrow openings called *gates*. This ensures that afterwards the whole "injection system" of runners and gates can be easily separated from the product (even so, the locations of gates remain visible on the product surface). For a product thickness of, say, 2 mm, gates are typically just 0.8 mm thick. Thanks to the very high shear strain-rate in the gate, the viscosity is relatively low at that point (see the far right corner of Figure 8.3), which is advantageous. However, there now is a potential problem and that is that the melt may shoot straight into the cavity. This phenomenon is known as *jetting* (Figure 8.4 left, middle) and leads to visible, ugly flow patterns, and brittle material. One common way to prevent this is to aim the melt against the wall at an angle, as also shown in Figure 8.4 (right).

Apart from this local issue, there is an important overall characteristic of proper injection. Many injection-molded products have a definite concave and convex side, or in other words, clear inside and outside surfaces. Injection is then usually done on the convex side (i.e., the outside). There are several reasons for this, which will become apparent in the course of this chapter.

---

**EXERCISE 8.15**

Determine for your sample products (Exercise 8.1) where the gates have been placed. How can you tell, and how can you know?

---

Pressure is normally the limiting factor in injection molding. So for larger products, it is often sensible to inject at multiple points and minimize the flow length $l$. Even so, there is a certain maximum ratio of flow length over thickness, $l/d$. In practice, this ratio should be below 100 for thicknesses of 0.5 mm, rising to around 400 for thicknesses of 2 mm or more, dependent (among other things) on the specific grade. One key consequence is that a large, thin-walled product can only be made using multi-point injection.

---

**EXERCISE 8.16**

Are there any disadvantages to multi-point injection?

---

**FIGURE 8.4**

Jetting (left, middle) and how to prevent it (right).

Once the melt has entered the mold through the gate(s), the actual mold filling begins. In this, it is essential to let the flow front expand gradually, much like you do when you pour pancake mix into a pan. This ensures that the air in the cavity is forced out completely and that the part contains no air bubbles. We can achieve this ideal by recognizing that the melt will always take the path of least resistance, so thickness and its variation over the product are crucial here. For instance, for smooth filling, the gates must be placed such that the melt flows "from thick to thin"—even better filling generally happens when the thickness is constant throughout the product.

---

**EXERCISE 8.17**

For a product that is essentially rotationally symmetrical, such as the bucket from Exercise 8.14, what are the two possible locations for single-point injection? Which is best, and why?

---

**EXERCISE 8.18**

Now imagine that to increase strength, the bucket wall thickness has been increased near the two handles. How would you then inject the product, and why?

---

If a product contains a hole, then the melt flows around that hole and merges beyond it. Where this merging takes place, we get a *knit line* (Figure 8.5). Such a line can often be clearly visible on the outside, spoiling the product appearance, and its mechanical properties can also be poor. So it is necessary to keep such lines out of sight and away from critical locations. Notice how multi-point injection also leads to knit lines. This second kind of knit line is known as "cold," whereas the former versions are known as "hot" or "warm."

---

**EXERCISE 8.19**

What measures can you take to influence the location where knit lines are formed, given a certain fixed product shape with several holes?

---

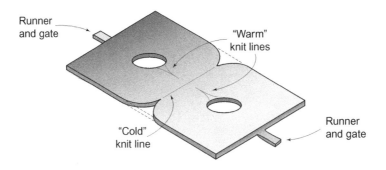

**FIGURE 8.5**

Formation of warm and cold knit lines near holes, due to multi-point injection.

Minimizing flow length, enabling gradual filling, ensuring easy removal of the injection system, and manipulating possible knit lines are not the only considerations to make when choosing the location of the gate(s). For instance, the mold layout may force you to inject from the side instead of the center of the product. As Section 8.7 will show, this happens when we have two-plate, multi-cavity molds. It is all too easy to end up with an uneasy compromise between function, cost, and quality!

When manufacturing parts of around 250 cm³ in volume, filling flows of some 300 cm³/s are normal. This volume flow rate gets higher for larger products and lower for smaller ones. So a midsize injection molding product such as the bucket from Exercise 8.15 is injected in just under one second. For the overall cycle time of injection molding, the injection stage is therefore of little relevance. For the cooling time, this is very different, as we shall now see.

## 8.5 COOLING AND EJECTING

The cooling time is a very important factor in injection molding, typically taking up 50% to 70% of the total cycle time for the process (less for thin-walled products, but more when using cheap molds with poor cooling facilities). Of course, the shorter this time, the faster we can manufacture and the sooner we will recover the investment for the molds. The cost of operating the molding machine itself also depends on the cooling time, because this cost is usually paid per unit of time. The cooling time itself depends strongly on the product geometry and so it is sensible to consider this relationship more closely.

A comparison with high pressure die casting of metals would seem to be in order (see Chapter 3). For this casting method, solidification and cooling times scale with volume $V$ over area $A$ (i.e., with linearly with part thickness $d$). But plastics and metals are very different: plastics have a two to four times higher specific heat capacity than metals and a 400 to 800 times (!) lower thermal conductivity. In other words: we can put relatively large amounts of thermal energy into plastics and this energy moves relatively slowly through the material toward the cold mold. So we cannot ignore the heat transfer inside the part (from core to surface) and should therefore expect that the part thickness influences the cooling time more strongly than it does for metals. (On the other side, there is no sharp phase transition from liquid to solid and the concept of latent heat does not apply: even for semi-crystalline plastics the extra enthalpy of crystallization is negligible in a first approximation.)

Once again, consider a flat channel with constant thickness $d$, but now fully filled with the melt. Initially, the melt processing temperature is $T_p$ and that of the mold wall is $T_d$ (this temperature is maintained using built-in cooling channels in the molds); the temperature at which we can eject is when the center-line falls below $T_e$, with $T_p > T_e > T_d$. The precise temperatures differ by kind and grade of plastic and depend as well as on the precise part geometry, but if they are known we can estimate the cooling time $t_{cool}$ as:

$$t_{cool} = \frac{d^2}{\pi^2 a} \ln\left(\frac{\pi}{4} \frac{T_p - T_d}{T_e - T_d}\right) \qquad (8.3)$$

where $a$ is thermal diffusivity (units: $m^2 s^{-1}$), defined as $a = \lambda/\rho c$, in which $\lambda$ is the thermal conductivity of the material in W/(m.K), $\rho$ is its density in kg/m³ and $c$ its specific heat capacity in J/(kg.K). According to this simplified formula, the cooling time only depends on the three temperatures $T_p$, $T_e$, and $T_d$, the thickness $d$, and the kind of plastic. Note the quadratic dependency on $d$, which is similar to

Chvorinov's rule for metal castings. Table 8.1 gives indicative data for several common plastics. Watch closely: the ejection temperature $T_e$ chosen is determined by the thickness of the product. Thin products force us to choose a $T_e$ on the low side of the range given in Table 8.1, because otherwise such parts have insufficient strength to resist the force of ejection. For thicker products, we can choose to eject at higher center-line temperatures because the plastic nearest to the mold will have cooled sufficiently to resist the ejection forces. Even so, this is not a rigorous rule, and in practice, $T_p$, $T_e$, and $T_d$ are always optimized through trial-and-error.

**Table 8.1** Indicative data for cooling time calculations for several common plastics

| Plastic | $T_p$ [°C] (range) | $T_d$ [°C] (range) | $T_e$ [°C] (range) | $a$ [× $10^{-9}$ m²/s] (average) | $\rho$ [× $10^3$ kg/m³] (average) |
|---|---|---|---|---|---|
| LDPE | 170–245 | 20–60 | 50–90 | 90 | 0.79 |
| HDPE | 200–300 | 40–60 | 60–110 | 75 | 0.82 |
| PP | 200–300 | 20–100 | 60–100 | 90 | 0.83 |
| PVC (hard) | 150–210 | 20–70 | 60–100 | 80 | 1.35 |
| PS | 160–280 | 10–80 | 60–100 | 75 | 1.01 |
| ABS | 200–270 | 50–80 | 60–100 | 75 | 1.03 |
| PLA | 150–200 | 10–30 | 20–50 | 75 | 1.10 |

---

**EXERCISE 8.20**

Calculate the cooling time for the cell phone from Exercise 8.9. What would this time be if we doubled the thickness? Answer the same questions for the bucket from Exercise 8.14.

---

**EXERCISE 8.21**

Revisit your answer to Exercise 8.12. Did you make the right prediction for what the drawback would be, and can you now also give a quantitative answer?

---

Of course, nearly all injection-molded products are more complex than a flat plate, with conse-quences for the cooling time. For instance, deep cores can increase this time significantly, especially if the cores are not equipped with cooling channels (there are separate, empirical formulae to estimate their influence on the cooling time, but such details do not concern us now). In practice, accurate pre-diction requires computer assistance, just as for injection pressures. However, in nearly all cases the cooling time depends strongly on the product thickness. And this takes us to the central dilemma of injection molding as a manufacturing principle: *the thicker the product, the easier its injection will be, but the longer its cooling time*—and vice versa!

Once the part has cooled down sufficiently, it can be taken out of the molds. This is done using ejection pins: these are integrated into the mold and lie flush with the surface of the cavity during injection and cooling, and these pins force the part out once the mold has opened. For practical reasons, they are almost always placed on the side opposite to the one from which the part is injected. Ejection is fast and does not significantly affect the cycle time, but the pins tend to leave visible marks on the material, and the ejection side should therefore preferably not be the visible side of the part. But note that because the gates will also leave marks when they are on the opposing injection side, this is not an easy problem to solve! Problems concerning ejection can always be solved, but at a price. Furthermore, consider that it is technically not possible to place a cooling channel and ejection pin at the same location, and that mold design therefore always involves a compromise between cooling and ejecting. All the more reason to consider these aspects well during design!

---

**EXERCISE 8.22**

Determine for your sample products (Exercise 8.1) where the ejection pins were located. Can you tell why have they been placed right there?

---

**EXERCISE 8.23**

Suggest three different ways to prevent the ejection from leaving visible marks on the part. For each, give one disadvantage also. Hint: see Equation (8.3) and Table 8.1.

---

Notice how the ejection pins can only do their job if the part remains attached to the ejection side of the machine once the mold opens. In other words, the part has to be drawn *away* from the injection side. This does not always happen automatically and is yet another concern for the designer. Section 8.7 explores this issue deeper.

---

## 8.6 SHRINKAGE, RESIDUAL STRESSES, AND VISCOELASTICITY

Compared to casting of metals such as aluminum or cast iron, the temperatures that occur during injection molding of plastics are low. So you would perhaps expect that shrinkage plays no significant role here. However, the thermal expansion of plastics is typically five to ten times that of metals, and cooling from processing temperature to room temperature (RT) therefore gives considerable shrinkage, despite the modest temperature change. Note that it is important to distinguish carefully between linear shrinkage (i.e., the parameter found in most data tables: the coefficient of linear thermal expansion, or CLTE) and volumetric shrinkage (which is three times greater). For amorphous plastics, volumetric shrinkage is typically some 1.5%; for semi-crystalline plastics, it can be up to 7%: the higher the degree of crystallinity, the larger this shrinkage will be. Wherever we have concentrations of material, we will then get unattractive "sink marks" on the outside of the part—for example, where we have reinforcing ribs or screw bosses on the inside. What is more, this type of shrinkage depends on the cooling speed: the slower we cool, the more opportunity the material has to crystallize (for good measure, we must add that different semi-crystalline plastic types can have very different crystallization speeds). And note that in the days following injection molding, this process can continue.

---

**EXERCISE 8.24**

Do you see any sink marks on your sample products (Exercise 8.1)? And what plastics are used in them: amorphous or semi-crystalline types?

---

To counteract shrinkage, we can exploit the fact that (unlike molten metals) plastic melts are somewhat compressible. If we "squeeze in" the material during injection, we force the polymer chains tightly together. Upon releasing the pressure, the chains spring back, negating (part of) the thermal shrinkage. (This is the "second reason" mentioned in Section 8.3 for applying more pressure than what is needed to overcome melt viscosity.) This tactic does, however, require considerable pressure: for instance, for PS, each 100 bar of pressure only reduces the volume by 0.3%. Another way to counteract shrinkage is to dimension the mold so that the product comes out precisely right after shrinkage, and this is also commonly done, although it demands complex simulations.

---

**EXERCISE 8.25**

How much extra pressure would we need to fully counteract shrinkage in PS? And how much in PP, assuming it has the same compressibility as PS?

---

The compressibility has a peculiar effect on the formation of internal residual stresses. In injection molded parts and products, these stresses are not primarily driven by temperature gradients, as they are in metal castings, but by the pressure history during filling and cooling—which varies from one location to the next and indeed, which even varies over the thickness. Consider that the material closest to the mold surface will "freeze" under the maximum pressure and hence sees the maximum loss of volume due to compressibility. The shrinkage that these outer layers undergo relieves pressure on the inner core and hence reduces the compression of the core (the falling temperature adds to this effect). In other words, over half the thickness, all material gets frozen at different pressure levels and consequently, with different internal stress levels. After cooling and ejecting, these stresses either distort the product or—if the geometry constrains this result—remain as residual internal stresses. In the latter case, the core can contain either slightly tensile or compressive stresses, depending on the thickness of the compressive sub-surface layer. Contrary to common opinion 10 years ago, thermal shrinkage effects are of lesser importance than the compressibility, when it comes to internal stresses.

Does the complexity end there? No, it does not, with significant consequences for design and manufacturing. A polymer melt combines viscous behavior (= a property of liquids) with elastic behavior (= a property of solids), giving "viscoelasticity." Not only does this cause the non-Newtonian behavior discussed earlier, it also leads to flow orientation and anisotropic material properties—not just in terms of the strength, strain-to-failure, and other mechanical properties, but also in terms of shrinkage. Viscoelasticity also affects the use and application of plastics, as it lies at the basis of the phenomenon known as "creep" (i.e., slow, continuous deformation under a static load, in materials loaded at temperatures above half their melting point, in Kelvin)—which is always the case for plastics. Think what happens to the handles of a plastic carrier bag if you put too much in the bag.

## 8.7  INJECTION MOLD DESIGN

This section presents the various types of molds that are in use today. For reasons of brevity, we limit the discussion to the main concepts, structured into four sub-sections: (1) layout, (2) parting plane, (3) number of cavities, and (4) special features. Throughout this discussion, it is best to consider a mold as a separate, complex machine that has to withstand large forces with negligible deformation—and that for a typical life span of 1 million "shots." The material of choice is tool steel (usually shaped using electro-discharge machining, but mechanical machining is also used) that will usually have received certain surface treatments, such as nitriding (see Chapter 13), polishing, or texturing. Aluminum molds also exist: these are cheaper but have a shorter life span. Note that not all of the mold actually needs to be considered as product-specific: the mold frame, plus part of the ejection system, is often standard and can sometimes be reused for other products.

1. For layout, we can classify injection molds into three main groups: two-plate molds, three-plate molds, and hot runner molds (Figure 8.6). Notice how the various plates are always linked using guiding pins and how there are two sides to the mold: the injection side and the ejection side. The two-plate mold is the simplest and potentially cheapest layout. Injection is done through the top plate straight into the product (for single-cavity molds) or via a runner system (for multi-cavity molds). One drawback of two-plate molds is that the injection system remains attached to the product and must be separated later. Another drawback, for multi-cavity molds, is that injection is done at the side of each product, which may not be ideal. The three-plate mold does not have these disadvantages: in this layout, the product cavity lies between the last two plates (as seen from the injection side) and most of the injection system between the first two. Once the mold opens, the product is automatically separated from the injection system. This layout also allows central injection in multi-cavity molds. Finally, in the hot runner layout, the injection system remains hot and does not need separation.

---

**EXERCISE 8.26**

Which layout is best if labor is cheap and investment must be minimal? And which is best if labor must be minimal and investment can be higher?

---

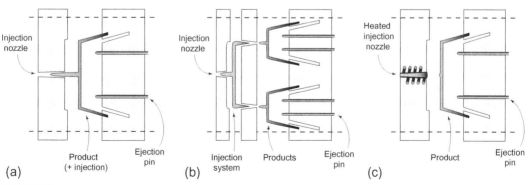

(a)     Product (+ injection)     Ejection pin     (b)     Injection system     Products     Ejection pin     (c)     Product     Ejection pin

**FIGURE 8.6**

Layout of (a) two-plate, (b) three-plate, and (c) hot runner molds. (For color version of this figure, the reader is referred to the online version of this chapter.)

---

**EXERCISE 8.27**

Which layout would minimize the formation of production scrap (i.e., material that is processed but does not end up in the actual part or product)?

---

2. Next we consider the parting plane. As Section 8.3 explained, the placement of this plane, together with the maximum pressure, determines the necessary clamping force and hence the size of the injection molding machine. But above all, it is important that the product (including the injection system) has a certain draft angle relative to this plane to allow ejection. Figure 8.7 shows how this can be done for the now familiar bucket. A draft angle of 3 degrees is usually sufficient, unless the product is deeply textured. Right where the parting plane meets the part, the part will get a distinct parting line. This is visible on the product as a surface defect, so consider this well when choosing the parting plane's location.

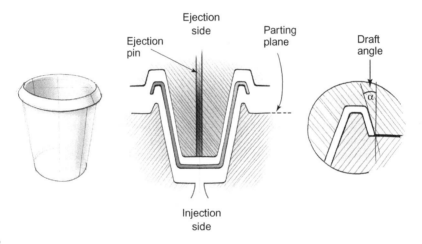

**FIGURE 8.7**

Mold halves, parting plane, and draft angle for an ordinary bucket.

---

**EXERCISE 8.28**

For your sample products (see Exercise 8.1), determine where the parting planes have been placed, and why. Also, determine if all draft angles are positive. If not (i.e., there are "undercuts"—places with negative draft angles), how have they been made?

---

3. Now we think about the number of cavities. For very high production volumes, injection molding with single-cavity molds is too slow. To speed up the process (and to deploy the machines more productively), it is then customary to use multi-cavity molds, with up to 64 identical cavities. Figure 8.8 shows an example. But speed is not the only consideration here. When a product is

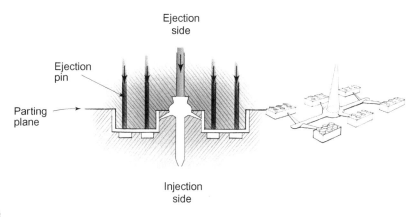

**FIGURE 8.8**

Six Lego bricks with central "submarine" injection.

relatively long and narrow, it does not effectively fill the mold surface, which is almost always rectangular. To make better use of the available area, two or more long products are placed side by side (Section 8.10 shows an example).

Notice also in Figure 8.8 how the injection system is deliberately shaped to have a small undercut, which ensures that it, together with the six products, remains attached to the ejection side upon mold opening: the big, central ejection pin has no difficulty overcoming this undercut. Also, look how the gates are placed just under the parting plane: this is called "submarine gating" or "tunnel gating" and allows automatic separation of products and injection system, even with a two-plate mold. Clearly, details matter!

**EXERCISE 8.29**

The set-up shown in Figure 8.8 will not produce identical products. This is because the two central blocks have a shorter runner than the other four and hence receive more pressure. How can we fix this problem?

**4.** In closing this section, we discuss four special mold features. First, there are *slides*. When it is not possible to get positive draft angles all around, even with smart parting plane placement, we can use a mold with parts that can move in directions other than the parting direction. Figure 8.9a shows this principle for an everyday office product: a plastic crate. Figure 8.9b shows the same solution in cross-section: notice how the slides move along inclined pins and are automatically pulled sideways once the mold opens. This is known as a "crate mold." It is a common solution for which standard mold frames are available, but the mold cost is considerably higher than for a comparable, simple "open-close mold." Also, the distance that a slide can move sideways is limited to around 20% of the mold stroke (i.e., the maximum distance between the mold halves). If this is not sufficient, then the slides can also be controlled with separate hydraulics, but that is even more expensive. Whatever the type, slides always leave slight marks on the product surface, like parting lines.

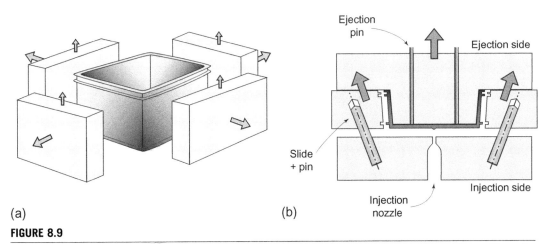

(a) (b)

**FIGURE 8.9**

(a) Schematic of the use of slides to mold a paper crate; (b) detail of crate mold with slides and pins. (For color version of this figure, the reader is referred to the online version of this chapter.)

---

**EXERCISE 8.30**

A standard snap-fit joint has a negative draft angle. Think of three different solutions to this problem. Which drawbacks do they have, and what is your preference?

---

A second special feature is the *stripper ring*. On thin, fragile products, ejection pins often cause unacceptable damage, even with comparatively low ejection temperatures. A stripper ring solves this problem by dividing the ejection pressure over the full product edge (see Exercise 8.24). Figure 8.10 shows this solution. Notice that for the product in question, the shrinkage will ensure that this part remains attached to the ejection side of the mold—the use of a small undercut as in Figure 8.8 is therefore not necessary.

Third, there is *mold venting*. During injection, the air must be able to escape from the cavity, and this requires special solutions. Usually, this takes the form of a small, thin channel placed flush with the parting plane, just at the end of the flow length. Note that ejection pins are never fully airtight and will also serve as vent holes. High form details such as ribs (that is, features that lie far above or below the parting plane) need separate vent holes.

The last special feature worthy of mention is *mold cooling*. Injection molds are always equipped with at least a few simple, straight channels for water cooling and temperature control, without which it is impossible to optimize manufacture. Also, by placing extra cooling channels at certain places we can often improve the dimensional accuracy. A relative newcomer is conformal cooling, where we have curved cooling channels that envelop the part, making cooling more homogeneous and improving accuracy. Still, even setting aside costs, mold design remains a compromise between cooling and ejecting: you cannot do both at the same spot.

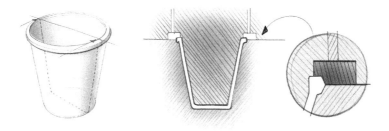

**FIGURE 8.10**

Stripper ring applied for a thin-walled bucket.

---

**EXERCISE 8.31**

Opening and closing of molds takes time (typically, the moving half moves at 1 to 2 cm/s). What does this imply for the cycle time of flat products as compared to deeper ones?

---

## 8.8 INJECTION MOLDING OF SPECIAL MATERIALS

As mentioned in the introduction, injection molding as a manufacturing principle is possible using all kinds of materials. This, however, does not mean that a designer can simply select any material, because the processability can be quite different for different materials, and not just in terms of the MFI. This section presents two examples to illustrate the possibilities and problems: (1) glass-fiber reinforced nylon and (2) PLA, a 'bio-based plastic'.

1. To increase the strength and stiffness of plastics, it is common to add short ($<$ 1 mm) pieces of glass fiber. Often, this is done for nylons, or polyamides: for instance, the designation "PA6GF30" stands for polyamide-6 with 30% glass fiber, by weight. Other plastics that are commonly glass-filled include ABS, PP, and PBT. This additive has two important consequences for processing. First, it may require up to twice the pressure to mold such grades as compared to unfilled ones, depending on the amount of glass filling. Second, the glass is highly erosive and limits the mold life span to 250,000 shots for hardened molds and just 100,000 shots for standard molds. So it is necessary to weigh the benefits and drawbacks carefully.

---

**EXERCISE 8.32**

Choose a common engineering plastic, such as ABS or PA, and look up what changes in mechanical properties (strength, stiffness, strain-to-failure, fracture toughness, etc.) can be made by the addition of glass fibers. (The CES software is particularly useful here.)

---

**EXERCISE 8.33**

Glass fibers have a lower specific heat capacity and a higher thermal conductivity than plastics. What does that imply for the cooling time for such grades: specifically, which term changes in Equation (8.3)?

---

2. Recently there has been renewed interest in plastics based on renewable resources as an alternative for fossil fuel based (crude oil) feedstock. Alternative base monomers for these *bioplastics* come from plants or algae, for example, polylactic acid (PLA). It is already being applied in all kinds of packaging, usually shaped by thermoforming or extrusion blow molding, but PLA can also be injection molded. Compared to commodity plastics such as PS and PE—arguably PLA's main competitors for low-cost applications—there are two interesting differences. First, PLA's viscosity is quite high and nearly independent of the shear strain-rate. Second, the amorphous grades of this plastic have very low shrinkage (less than 0.5% during the full process). In injection molding, the high viscosity is a drawback, whereas the low shrinkage is a benefit. This appears to be a pattern for bioplastics: overall their processability is not better or worse, but simply different from that of non-renewable plastics. Likewise, their properties often present an interesting new mix, and one-to-one replacement of plastics by bio-based alternatives (called "drop-in" by industry) requires careful consideration. Good design is therefore essential for these materials to gain market share. Note that bio-based does not necessarily imply biodegradability: bioplastics can be just as durable as conventional plastics.

---

**EXERCISE 8.34**

Look up the key mechanical and thermal properties of PLA. Also, consider its price. How do these values compare to those of PS or PE?

---

**EXERCISE 8.35**

In injection molding of PLA, the injection time is typically three times longer than that of most comparable plastics. Gate design is also different. Explain why, with reference to the theory covered in Sections 8.3 and 8.4.

---

**EXERCISE 8.36**

Is PLA a sustainable material? Explain your judgment.

---

In closing, it bears pointing out that PLA is just one of the many bioplastics available and that the properties of PLA alone vary considerably across its various grades, just as with other plastics. This variety keeps increasing, leaving little doubt that bioplastics will become only more important in the near future.

## 8.9 MANUFACTURING METHODS FOR INJECTION MOLDING

So far, we only discussed injection molding as a manufacturing principle and the material we covered is universally valid in principle. Now it is time to discuss several common manufacturing methods. Of the many methods (there are nearly a hundred!) we present these five: compact injection molding, insert/ outsert molding, injection molding using gas injection, injection molding with in-mold decoration (IMD), and 2K injection molding.

### Compact (or standard) injection molding

This is the standard method for injection molding, used in an estimated 80% of all injection molded parts and products. One "shot" of molten plastic is injected into a cavity (or multiple cavities), cooled, and ejected. The method dates back to the 1920s but really took off only around the 1950s with the advent of standardized machines, in parallel with the rise of mass consumption (which it arguably helped to create). It is best suited to high production volumes of complex, 3D-shaped products. However, it can also be a solution for smaller volumes, when the form freedom outweighs the high depreciation of the mold cost per part. Machines for compact molding are mainly classified by tonnage, meaning the clamping force they can produce, but the maximum shot volume (i.e., the amount of material that the machine can inject) is a second notable indicator. Machines typically range between 25 tons of clamping force to more than 2,500 tons.

---

**EXERCISE 8.37**

Give three guidelines for the design of "standard" injection molded products. Each time, make explicit reference to the theory covered in the preceding sections. How are your guidelines reflected in your sample products (Exercise 8.1)?

---

### Insert/outsert molding

In this method, specific parts or components are placed inside the cavity prior to molding, and during injection, the melt flows around these "inserts." This allows integration of, for instance, metal strips into the product. If the part to be integrated is larger than the actual plastic elements and acts, so to speak, as the product backbone, then it is referred to as "outsert molding." In both instances, the joint between plastic and insert/outsert is qualitatively very good. Machines for this method have their injection unit oriented vertically, aiming downward, while the mold is oriented to have its parting plan horizontally. This setup, which is different from normal injection molding, helps keep the inserts/outserts in place prior to injection.

---

**EXERCISE 8.38**

Differences in thermal expansion between insert and plastic may cause problems. Calculate this difference for a typical injection molding temperature for a flat, straight 0.5 mm thick steel strip placed flush onto a 1.5 mm nylon plate. What exactly will be the problem?

---

## Injection molding using gas injection

In this method, a gas (usually nitrogen) is injected into the product in those locations where internal cavities are needed. This is done using a thin needle, just after the melt is injected and partially fills the cavity. Gas pressures are typically around 300 bar. Essentially, the melt is blown up like a balloon until it fills the mold completely and begins to cool down. Then, the gas pressure is released and the needle is retracted. The method prevents the use of expensive sliding cores and, particularly for larger products, allows a large reduction of part thickness. This shortens the cycle time considerably; as a bonus, inaccuracies due to large shrinkage are also reduced. Because it is difficult to keep the gas bubble in place during cooling, the method is only applicable for relatively simple shapes, such as handles and pipes. A still largely experimental variety is to use water injection.

---

**EXERCISE 8.39**

Using a cross-sectional view, explain why a gas bubble will most likely not remain in the center of a non-symmetric part. Hint: melt viscosity goes up where the melt cools down. What kind of design rule does this suggest?

---

## Injection molding with in-mold decoration (IMD)

IMD is a method in which, prior to plastic injection, a thin decorative skin is placed inside the cavity. This skin has been made prior to molding, using thermoforming. Alternatively, thin foils can also be used, which are shaped during injection itself through the process pressure. During the process, the melt attaches itself to the decoration, giving a high-quality finish. This way, ordinary plastics can acquire the look of wood grains, carbon fiber, or even marble. By applying a different skin or foil, the part can also be changed at relatively low cost. Every modern car interior is full of IMD applications, but the method is also gaining popularity outside the automotive industry.

---

**EXERCISE 8.40**

On which side of the part do you get the decoration: injection or ejection side? And what does that mean for the suitability of IMD in practice?

---

## 2K injection molding

This method allows the injection of more than one shot of different plastics into the same mold to create one integrated product. Each component is injected by a separate injection unit. "2K" is the most common (from the German "2 Komponente"), but more are certainly possible: the current record is 7K! The method allows the combination of different materials (e.g., PP and EPDM, as presented in Section 8.10) or just different colors (e.g., keyboards) that can either be joined tightly, or have the freedom to move relative to each other as a hinge. "Sandwich molding" is yet a third possibility: here, one material lies fully enclosed inside another (e.g., virgin plastic outside, recycled plastic inside). There are three common mold solutions to enable 2K molding: the use of sliding or translating cores, the use of a rotating "table" inside the mold, and the use of a pick-and-place robot. In the first solution, the mold does not open during the process and injection of the various components is done sequentially, increasing the cycle time. In the last two solutions, the mold does open during the process and the components are injected simultaneously (i.e., in effect, the mold is of the multi-cavity type). The method is expensive in terms of investments but allows huge integration of functions.

---

**EXERCISE 8.41**

Which of these five methods can be combined with one another? Also, think of a realistic application for each combination.

---

## 8.10 A WORKED-OUT COST EXAMPLE

This section takes a closer look at the cost of injection molded parts and products, doing so for the case study of a "Swiffer": a handy household item for cleaning windows (Figure 8.11). The Swiffer as shown was produced using 2K molding, with a PP handle (white) and EPDM blade (gray). This second material is a thermoplastic elastomer, a relatively new type of polymer that is well suited for injection molding. Here we will analyze the manufacturing cost of this product by examining how the production volume and the country of manufacture influence the result. Also, we will ask the question, if "2K" was the proper choice: is it cheaper than separate molding of the two parts followed by assembly, or not?

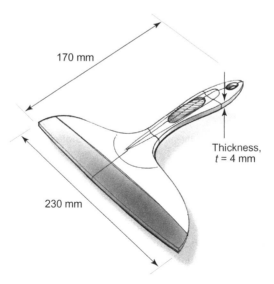

170 mm

Thickness, $t = 4$ mm

230 mm

**FIGURE 8.11**

2K Swiffer.

*(Design: Pilipili, Kortrijk, Belgium.)*

---

**EXERCISE 8.42**

Make a first, rough estimate for the manufactured cost of a 2K Swiffer. (Hint: due to taxes, margins, sales costs, and so on, the sales price for products of this type is typically three to five times the manufactured cost.)

---

Manufacturing costs can be broken down in many different ways. A simple and common approach for mass-produced parts is to state the cost per part $C_{\text{part}}$ of manufacturing $N$ products as:

$$C_{\text{part}} = C_m + \frac{C_C}{N} + \frac{\dot{C}_L}{\dot{N}} + C_t \tag{8.4}$$

Here, $C_m$ covers the costs of material (and other consumables, such as packaging) in each part, $C_C$ represents the dedicated investments needed to make the component (tools, molds, dies, etc., but not the machine itself: see below), $\dot{C}_L$ incorporates all costs that are accumulated at an hourly rate, and $\dot{N}$ is the rate of production. Note that there are two quite different contributions to $\dot{C}_L$: (1) conventional hourly costs, such as labor, energy, and so on, and (2) the major capital machine investments that are amortized over a long write-off time, such as five years, and thus converted into an equivalent hourly cost. In addition, we consider a further cost per part $C_t$ covering transport costs (with a number of assumptions, detailed later). We can apply this model to both the 2K Swiffer and a "1K alternative"—we have the same six cost parameters, but simply with different values.

1. First we consider the material costs. The 2K Swiffer weighs 50 grams: 30 grams of PP and 20 grams of EPDM. For injection molding, we of course need a bit more, but for simplicity we assume the injection system is fully recycled at zero cost (alternatively, we can assume the molds to have "hot runners"). The cost of plastics, especially for the cheaper ones, depends strongly on the oil price. The location, however, is of little influence: prices of commodity polymers are uniform around the world. We will also neglect the cost of packaging.

---

**EXERCISE 8.43**

Look up the costs of PP and EPDM. What is the material cost per 2K Swiffer? Is it reasonable to assume the same material cost for the 1K alternative? How does it compare with the estimate made in Exercise 8.42?

---

2. Next we consider the mold costs. As noted in the introduction, these costs vary strongly worldwide. However, if we make certain requirements regarding mold life span and product accuracy, the variation gets considerably smaller. For this case study we can assume for the 2K mold a cost of €35,000. This will be a single-cavity mold with a sliding core (so sequential injection of the PP and EPDM) with a guaranteed life span of a million shots. For the 1K Swiffer, we need two molds: assuming both molds have the same life span as that for the 2K mold, the single-cavity mold for the handle costs €20,000, and the two-cavity mold for the blade costs €18,000. Notice that the blade is long and narrow, and for such shapes a single-cavity mold is not efficient (see Section 8.7).

---

For both the 2K Swiffer and the 1K alternative, what is the capital cost of the molds, per part, when shared over production runs $N$ of (a) 100,000, (b) 1 million, and (c) 10 million?

---

3. To determine the labor and other hourly costs per part, we first must know the cycle time. For the 2K Swiffer, we can assume this to be 80 seconds per shot; for the 1K Swiffer, we assume 40 seconds for the handle and 40 seconds for the blade (two parts/shot, for an effective cycle time of 20 seconds), plus 6 seconds for manual assembly (a one-person job going on in parallel with the injection molding). This gives us an hourly production rate for each variant. Next we consider the hourly labor costs. Injection molding machines can work fully automatically, but some kind of supervision will be necessary, with one worker controlling a certain number of machines, which will also vary from country to country. Table 8.2 gives some realistic data—using this data we can evaluate the labor costs per hour in three regions.

**Table 8.2** Labor and Machine Costs in Euros per Hour in Three Regions

|  | The Netherlands | Czech Republic | China |
|---|---|---|---|
| Labor costs (two shifts) | 19.30 | 11.70 | 4.90 |
| Number of machines overseen by one worker | 6 | 3 | 2 |
| Costs 2K machine, 140 tons | 32.50 | 28.50 | 25.80 |
| Costs standard machine, 40 tons | 15.00 | 13.00 | 11.00 |
| Costs standard machine, 80 tons | 20.00 | 18.00 | 16.00 |

As Table 8.2 shows, the personnel costs vary strongly per region, which is obviously a strong driver for moving production offshore. The costs are given for manufacture in two shifts, which is usually the optimum for this kind of production, with typically around 3,000 effective production hours per year. One shift would mean that the invested capital is not used efficiently, whereas "24/7 production" would require five crews, plus high premiums for working on weekends and on holidays. Of course, especially in low-wage countries such as China, actual production hours are longer but productivity tends to be poor as compared to the situation in fully industrialized countries, and effectively, the hours are comparable.

---

What are the labor costs per hour, and per Swiffer, for both varieties? Determine these costs for all three regions mentioned.

---

Table 8.2 also gives data on the machine costs, converted to an equivalent hourly rate. These costs combine elements such as the power consumption, maintenance, and depreciation of the machine itself, plus the overhead costs for the factory space it requires. Machine costs mainly depend on machine size and functionality, and to a lesser extent on geographical location. We will assume that the 2K Swiffer requires a 140-ton 2K machine, whereas the 1K Swiffer needs a 40-ton standard machine for the handle and an 80-ton standard machine for the blade (due to the two-cavity mold).

---

**EXERCISE 8.46**

Determine the machine costs per hour, and per Swiffer (both varieties, all three regions).

---

4. Finally, for transport costs, we first assume that the Swiffers are intended for the Western European market and are distributed from Rotterdam. Next, we assume manufacture can take place close to the market itself (in the Netherlands), in an Eastern European country such as the Czech Republic or in a Far Eastern country such as China. Local manufacture means we can ignore transport costs. Manufacture in the Czech Republic involves approximately €1,000 worth of transport over land to Rotterdam per 12 meter container, and manufacture in China involves roughly €2,000 worth for transport overseas, also per 12 meter container. Such containers can hold up to 25 tons of goods, although for comparatively low-density goods such as packaged Swiffers they will be full long before the weight limit is reached.

---

**EXERCISE 8.47**

Estimate the number of packaged Swiffers that can be transported in a full 12 meter long container (with a cross-sectional area of 2 × 3 meters). Hence, estimate the transport cost per Swiffer for manufacture in the Czech Republic and in China (for batches of 100,000, 1 million, and 10 million). What other disadvantage is there to offshore manufacturing?

---

**EXERCISE 8.48**

Now, add up cost components (1)–(4) to determine the total costs of both Swiffers for all three regions (for production runs of 100,000, 1 million, and 10 million). What is cheaper: 2K or 1K? Which region is "best"? Also, compare your answer to the estimate you made in Exercise 8.42.

---

Of course, this analysis only gives an approximation, but it is still instructive and useful during conceptual design. In principle, the approach can be applied to all manufacturing processes. All you need is good data, common sense, and some experience. More detailed analyses, however, are a specialist's job and require much more time and effort.

---

**EXERCISE 8.49**

Time for a reality check! Use Sections 8.3 and 8.5 to determine whether the cycle times and machine tonnages assumed in the analysis are realistic, at least for the 2K Swiffer.

In closing, we should pay attention to the fact that several other cost components have not been considered, notably worker safety and environmental concerns, but also time-to-market (needed for product design and manufacturing engineering). Likewise, we have ignored any possible quality differences between the 1K and 2K Swiffers. These can be substantial: for instance, the smooth joint between handle and blade for the 2K variety quite likely delivers better performance than any joint we can manually make in the 1K variety in just six seconds. Taking such possibilities into proper account is key to good design!

## 8.11 SAMPLE PRODUCTS

The chapter has presented several examples of injection-molded products. What those examples have in common is that they are relatively simple objects, and although molding them can be highly optimized, it involves few real manufacturing challenges. This section presents eight products that, each in its own way, are more challenging. Together, they give further insight into what can be done using injection molding—arguably the most important manufacturing process for design. The discussion also gives the typical costs, investments, and production speeds for these sample products, assuming that manufacture takes place in highly industrialized regions and is handled by highly qualified suppliers. All data are approximate and representative for the year 2013.

### Lamppost cover

Figure 8.12a shows a lamp cover for a lamppost. This part must be vandalism-proof, making a high-transparency grade of polycarbonate (PC) the logical choice. It is 450 mm long and, between the many thin ridges, 5 mm thick. Apart from the extensive drying of the base granulate, manufacture presents a second challenge: mold detail and finish. The lamppost must generate evenly distributed light, but because the lamp element does not emit such light, the cover must act as a correcting lens. Consequently, its mold has an intricate, detailed pattern of lens ridges that can only be made on a state-of-the-art CNC-controlled multi-axis high speed milling machine (Figure 8.12b). Naturally, the mold surface must also be very smooth, so all those fine details must be highly polished as well. The mold has a single cavity and central injection using a needle valve. Transport is a third challenge: to get this part to final assembly without any scratches, a special packaging tray is required, made from expanded polypropylene (EPP). This part is also injection molded but can be reused.

- Material: "crystal clear" high-impact transparent PC
- Production volume: 25,000 units/year
- Part-specific investments: single-cavity mold 80,000 euro, plus 10,000 euro for the EPP packaging tray mold
- Cycle time: 60 sec = 60 units/hour
- Integral part cost: 8 euro

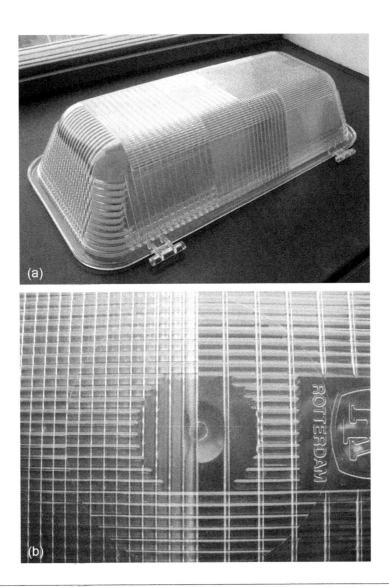

**FIGURE 8.12**

(a) Lamppost lamp cover; (b) injection detail. (For color version of this figure, the reader is referred to the online version of this chapter.)

## Camper toilet bowl

In Figure 8.13a, you see the toilet bowl used in various camper vehicles. It is 500 mm wide with a 3-mm wall thickness. For such parts, highly crystalline grades of polypropylene (C-PP) are the right choice: affordable, yet resistant to the aggressive cleaning fluids used in toilets, and available in crystal-white colors. However, such grades show considerable shrinkage during molding, and shaping it into a closely toleranced part such as this bowl requires advanced process simulation. Furthermore, considerations of hygiene require the surface to be very smooth, without any surface defects where dirt could accumulate.

**FIGURE 8.13**

(a) Camper toilet bowl (upside-down view); (b) injection detail. (For color version of this figure, the reader is referred to the online version of this chapter.)

Consequently, the marks from the centralized "film injection," placed right at the narrowest point of the bowl (see Figure 8.13b), must be cut off carefully, leaving a very smooth surface. Finally, this part has such deep undercuts that a standard crate mold would not work, as the distance that the slides must travel sideways is too far compared to the part's height. Instead, the mold has hydraulically operated slides.

- Material: C-PP, "crystal white"
- Production volume: 80,000 units/year
- Part-specific investments: single-cavity mold with hydraulic slides 100,000 euro, plus automated unit for removing the film injection 60,000 euro
- Cycle time: 45 sec = 80 units/hour
- Integral part cost: 5 euro/unit

## Nylon clothing tags

In virtually every piece of clothing sold in our shops, a tiny plastic tag is used to attach the price label. You would perhaps not expect it, but the base shapes of these ubiquitous items are also injection molded, after which they are heated with infrared light and stretched by 300% to obtain their final shape and tenacity. The material is nylon (PA66), which is not only easily molded but also lends itself well to heating and stretching. Production volumes are huge, and consequently, a mold with 200 cavities is used, with centralized injection onto the parting plane and a complex, well-balanced system of runners

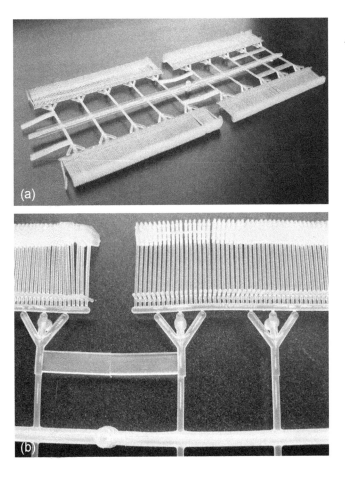

**FIGURE 8.14**

(a) Nylon clothing tags (as molded and stretched); (b) injection detail.

(Figure 8.14). Manufacturing this mold, with its numerous, closely toleranced details that must be able to withstand the high injection pressures, is challenging.

- Material: PA66
- Production volume: > 1 million units/year
- Part-specific investments: 200-cavity mold 100,000 euro, plus infrared heating-stretching unit 150,000 euro (reusable for other tag types)
- Cycle time: 24 sec = 150 shots/hour = 30,000 units/hour
- Integral part cost: < 0.01 euro/unit

## Connector plug

Our information technology–dominated world cannot exist without computer cables, and cables cannot exist without connector plugs. Figure 8.15a shows such a plug, connecting 64 individual wires, with Figure 8.15b providing a close-up. Here, the main manufacturing challenge is making the highly detailed mold with close tolerances (< 0.05 mm) to provide a reliable connection and give the right "click".

(a)    (b)

**FIGURE 8.15**

(a) Computer connecting plug; (b) close-up view. (For color version of this figure, the reader is referred to the online version of this chapter.)

Predicting shrinkage is a second challenge. Initially, this forced a choice for the (amorphous) ABS, but process simulations are sufficiently developed to allow a switch to the cheaper (semi-crystalline) PP, correcting the mold shape for shrinkage where required. Note that plugs require two parts. The data below are for the "male" part, but those for the "female" part are comparable.

- Material: PP
- Production volume: >100,000 units/year
- Part-specific investments: single-cavity mold (male part) 80,000 euro
- Cycle time: 45 sec = 80 shots/hour
- Integral part cost: 3 euro/unit

## Food tray with in-mold labeling

Many types of thin-walled packaging are thermoformed, but increasingly, these applications are being replaced by injection-molded parts. Figure 8.16a shows an example: a food tray. The main challenge

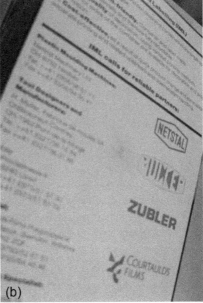

(a)  (b)

**FIGURE 8.16**

(a) Food tray; (b) label used in tray. (For color version of this figure, the reader is referred to the online version of this chapter.)

here is to minimize costs by making the part as thin as possible. By using high-flow grades of PP, this part can be made just 0.4- to 0.6-mm thick without the injection pressure getting too high, even with a single, centralized injection point. A second challenge is the use of in-mold decoration (Figure 8.16b). For such parts, foils are used that are shaped during the injection molding process, eliminating the need for preforming. Also, injection must be done on the tray's inside (i.e., on the opposite side of the foil).

- Material: PP, high-flow grade (MFI = 30)
- Production volume: 1 million units/year
- Part-specific investments: single-cavity mold 50,000 euro, plus in-mold labeling unit (integrated with mold) 60,000 euro, plus stacking unit 10,000 euro
- Cycle time: 8 sec = 450 units/hour
- Integral part cost: 0.13 euro/unit (including 0.04 euro for the label)

Note that investments such as the IML unit and stacking unit can in principle be reused for other products, but for very high production volumes they are best seen as product-specific.

## Front covers for domestic heating unit

Most people only see the heating unit in their house when it breaks down, and when that happens, they are not amused. So manufacturers spare few expenses to give their products the right "look and feel" so they can, in a way, save face, and a good front cover plate goes a long way to do that. Figure 8.17 shows the plate for one specific heating unit. It measures 300 by 150 mm and is 3 mm thick. To give this "eye-catcher" part

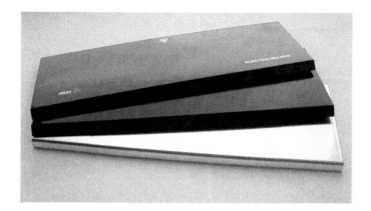

**FIGURE 8.17**

Front cover plates for domestic heating unit. (For color version of this figure, the reader is referred to the online version of this chapter.)

the right finish, a deep black, high-gloss, scratch-resistant label is molded onto the base plate material using in-mold labeling. This gives far superior surface quality than molding only with ABS (or other low-cost plastics)! Alternatively, a silver metallic surface can be created simply by using a different label (see Figure 8.17, made anonymous at the manufacturer's request).

- Material: ABS
- Production volume: 50,000 units/year
- Part-specific investments: single-cavity mold 40,000 euro, plus in-mold labeling unit (fully integrated with mold) 40,000 euro
- Cycle time: 45 sec = 80 units/hour
- Integral part cost: 4 euro/unit

## Truck air inlet filter

In 1999, a terrible fire raged through the Mont Blanc tunnel, killing 39 people. Subsequent investigations revealed that the fire probably started when a burning cigarette was tossed from a vehicle and got sucked into a truck air inlet, setting the paper air filter alight and spreading from there. Consequently, all trucks now must have additional mesh-type air inlet filters. These parts consist of a stainless steel wire mesh around a plastic frame and are suitable for retrofitting. Figure 8.18 shows one such filter, measuring 600 by 100 mm. It is made by insert molding (the mesh is the insert) with a PP edge and several diagonal stiffeners, followed by molding a flexible EPDM edge around it. So this part combines 2K molding with insert molding.

- Materials: PP + EPDM + stainless steel wire mesh
- Production volume: 30,000 units/year
- Part-specific investments: single-cavity 2K mold with 180-degree rotating table 70,000 euro, plus robotic gripper for mesh placement 10,000 euro
- Cycle time: 50 sec = 72 units/hour (note that injection of both components is done in parallel)
- Integral part cost: 6 euro/unit (including 2 euro for the wire mesh)

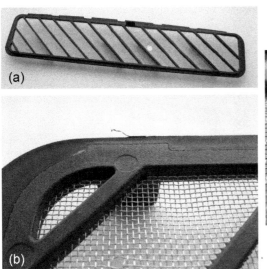

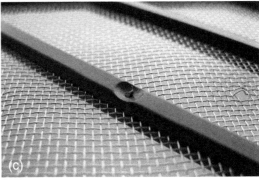

**FIGURE 8.18**

(a) Truck air inlet filter; (b) PP-EPDM edge detail; (c) stiffener detail plus injection point. (For color version of this figure, the reader is referred to the online version of this chapter.)

## Automotive air bag cover

Few products undergo such intense scrutiny as car steering wheels, and consequently, the air bag covers placed on them must simply look and feel perfect. Basically, these parts are hollow shells with snap-fit joints around the edges, a pattern of closely tolerated "tear lines" to allow proper air bag deployment, and a separate logo. The numerous manufacturing challenges begin with the shell itself, which is—for the air bag cover in question—made from a heavily rubber-modified grade of C-PP that is difficult to process due to high shrinkage and strong melt flow orientation. Also, it must have a special "soft touch" top layer, which is made by over-molding with SEBS, making it a 2K product. In addition, it is finished with a layer of scratch-resistant paint for even better surface quality. Figure 8.19a,b shows this component (made anonymous at the manufacturer's request).

The logo consists of two parts: a base plate made of a PC-ABS blend and the actual logo top that is made of ABS and chrome electroplated (note that this requires a special grade of ABS). This part is shown in Figure 8.19c. After plating, the logo is used as an insert in molding the base plate. The combination is then joined to the air bag cover by means of hot plate welding; the cheaper ultrasonic welding method would not give sufficient joint strength to keep the logo in place during air bag deployment. A benefit of using separate logos is that the same cover can be used for various brands. Given the increasing reliance on "product platform design," with several different car brands using the same body and parts, this is a major advantage.

A final challenge to the manufacturer of the air bag cover is the requirement to store all relevant process parameters for a 10-year period, so that in case of legal claims the carmaker can verify that the product was manufactured properly (or, if not, transfer those claims to the molder). These data are not

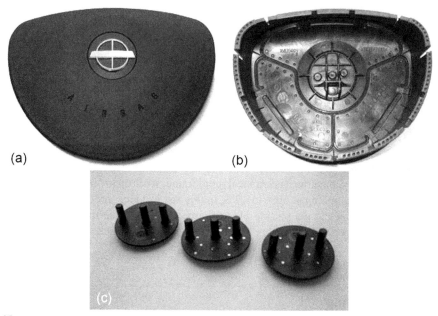

**FIGURE 8.19**

Airbag cover: (a) front view; (b) rear view; (c) logo inserts (rear view). (For color version of this figure, the reader is referred to the online version of this chapter.)

simply stored per production run, but per actual shot. All in all, injection molding such parts can be considered the "champion's league" of plastic manufacture!

- Materials: heavily modified C-PP and SEBS for cover, PC-ABS and ABS for logo
- Production volume: > 500,000 units/year (with different logos)
- Part-specific investments: single-cavity 2K mold with 180-degree rotating table for cover 200,000 euro, plus four-cavity mold for logo top (including automated placement unit) 75,000 euro, plus racks for chrome plating 5,000 euro, plus four-cavity mold for logo base plate 25,000, plus heat welding setup 50,000 euro, plus single-cavity mold for expanded PP cover packaging crate 25,000 euro, plus investment for automated processing data storage 25,000 euro, plus additional manufacturing engineering costs 100,000 euro—for a total of approximately 500,000 euro!
- Integral part cost: 5 euro/unit

## Summary

The challenges in injection molding can come in many forms. First, the material: grades that show strong shrinkage or melt flow orientation, as well as those that require extensive drying or high processing temperatures, are all relatively difficult to process. Shape is, of course, a second factor, with details such as large undercuts and extreme ratios of flow length to thickness complicating the operation. Next are the challenges inherent to complex methods, such as 2K molding, IMD, or insert molding, especially when these methods are to be combined. Quality considerations (essentially, all five attributes in the manufacturing triangle) pose a final challenge. Today's manufacturing industry can successfully tackle them, but clearly at some cost: just look again at the investment levels for the eight products described in this section! When considering the trade-offs among function, cost, and quality, it is good to know that when all is said and done, one combined element above all determines success: *communication and organization during the whole design project.*

Sources and tools that deal with mold costs and their estimations are in abundance, but not all are good. To separate the wheat from the chaff, you must first understand how mold cost and quality are related for a given product shape, size, and material. For example, suppose we put the price level for a mold made in Germany at 100%, guaranteed to have the legendary German "gründlich" quality level and life span. Then a mold made in China (for the same product, of course) would cost ~70%, assuming it is built to comply with European quality standards and DME or HASCO exchangeable parts. Still assuming European quality, but now using a local mold frame, we drop to 50% to 60%, and if we go for local quality standards, we end up at 30% to 40%. In this last instance, product quality will be visibly poor: more pronounced lines at the parting plane, warped products due to poor cooling, and so on. What is more, these quality problems will only get larger with time, because mold life span is generally poor as well.

## 8.12 Further Reading

As in the other chapters, this one has limited itself to the essentials. Terms such as *family molds* and *stack molds* have been left out, only five of the numerous methods have been presented, and several valuable theoretical concepts have been left unexplored. Likewise, our discussion of special plastics has been modest, ignoring high-end materials with spectacular performance, such as glass-filled polyacrylamide (trade name: IXEF 1022) or carbon-fiber-filled polyetheretherketone (PEEK). Furthermore, we have left injection molding of thermosets uncovered: it is done, but only for niche applications. If you wish to know more, virtually every book about manufacturing technology covers injection molding, and other suggested readings are listed here.

Various scientific journals address the many aspects of injection molding, such as *Polymer Engineering & Science.* As for accessible industrial journals, there are the German journals *Plastverarbeiter* (www.plastverarbeiter.de) and *Kunststoffe* (www.kunststoffe.de)—the latter being accessible in English. The largest European industry fair is 'K in Dusseldorf', held every three years (see www.k-online.de). And to explore the world of bio-based plastic, visit www.biopolymer.net.

Ashby, M.F., Shercliff, H.R., Cebon, D., 2013. Materials: Engineering, Science, Processing and Design, third ed. Butterworth-Heinemann, Oxford, UK. (*An introductory text at a complementary level to this book, providing detail on material properties and processing in a design-led context. Chapters 4 and 6 discuss the physical origins of Young's modulus and yield strength of polymers, and the effect of temperature on properties. Chapter 18 provides more detail on the cost model used in this chapter.*)

Hudson, J., 2011. Process–50 Product Designs from Concept to Manufacture. Lawrence King Publishing, London. (*Puts injection molding into context beautifully. Somewhat slanted toward "design as art," but still very much recommended.*)

Osswald, T.A., Menges, G., 2003. Materials Science of Polymers for Engineers, second ed. Hanser, Munich, Germany. (*Ideal in-depth reading material on virtually all aspects of the subject, including the various equations used in this chapter. Comparatively recent and highly recommended for those who want to dig deeper.*)

# Thermoforming

Thermoforming is often used to make packaging, with this disposable "clamshell" box for tomatoes being one of countless examples. However, the process is also very well suited for durable applications, from prototyping to large production volumes.

## 9.1 INTRODUCTION

As a manufacturing principle, thermoforming is relatively simple. A thermoplastic sheet or foil is heated, formed using a pressure difference into (or over) a cold mold, then it cools down and the edge is cut off. Thermoforming is an open process with relatively low forces. It is also quite recent: as the first application, specialists generally recognize the transparent acrylic domes of World War II aircraft. The product-specific investments are considerably lower than, for example, those related to injection molding. For this reason, thermoforming products can often be developed in a short time. The design freedom in the direction of the part thickness may be limited, but the principle is flexible with respect to production volumes; it also offers a substantial range of texturing and finishing options. Furthermore, thermoforming can also be used to process laminated materials, such as a thin, high-quality acrylic foil pressed onto a cheaper vinyl base. So, all in all, there are more than enough reasons to learn more about this process!

To say a bit more about production volumes: on one side of the spectrum, thermoforming is applicable to one-off manufacture and prototyping; on the other side, it can produce volumes up to hundreds of *millions* per year, as, for example, in disposable packaging or coffee cups. With such a wide range, it is obviously not possible to state *the* investment cost for the principle, also because—just as with injection molding—the prices for seemingly the same mold can vary considerably worldwide. However, the manufacturing principle is straightforward, the number of different manufacturing methods that are commonly used is small, and it is relatively easy to get a decent overview of what the manufacturing triangle looks like for this process.

One manufacturing principle that is closely related to thermoforming is *extrusion blow molding*. Here, the input material is not a flat sheet or foil, but an open or closed extruded tube, often with a pre-formed end. This preform tube, or *parison*, is then expanded into a mold. Countless plastic bottles and containers are made this way, with the threaded end already pre-shaped for the closure. Extrusion blow molding can be partly explained along the same lines as thermoforming, but for the details we nevertheless refer to the specific literature. For a discussion of plastic extrusion, refer to Chapter 13.

---

**EXERCISE 9.1**

Search around your house for examples of thermoformed and blow-molded parts or products. Hint: don't forget to look at packaging applications.

---

## 9.2 RECAP: THERMOPLASTIC BEHAVIOR

In our discussion of injection molding we have seen that above the glass transition temperature ($T_g$), thermoplastics become rubbery and moldable. This behavior lies at the heart of the thermoforming process. As we have also seen earlier, the flow response above $T_g$ differs widely depending on the plastic (e.g., LDPE, PS), but often also for different *grades* of the same plastic (e.g., standard PS, high-impact PS, etc.). Some have an abrupt glass transition: even a small increase in temperature is sufficient to cause a significant decrease in the stiffness. This is the preferred behavior for thermoforming, because the whole process can then take place within a narrow temperature range, which allows for a high productivity. We also know that at a higher softening temperature ($T_s$), the plastic softens and becomes viscous, again depending on the plastic type and grade. Injection molding is done above $T_s$, but in thermoforming we should not exceed this value. It is also favorable if the material has a large difference between $T_g$ and $T_s$. The "ideal" thermoforming material has a stiffness (or viscosity) that is more or less constant in this range, because we do not then need to control the temperature accurately. Examples are high-molecular-weight grades of amorphous plastics such as PS, ABS, PVC, PMMA, and PC: all are plastics with a relatively long, horizontal rubbery "plateau" in the graph of stiffness versus temperature. All of these plastics are commonly used in thermoforming.

Other plastics have a more gradual glass transition and only a short rubbery plateau, or even a stiffness response that drops continuously with increasing temperature and never levels off. To thermoform such materials, we need to heat them to temperatures far above the $T_g$ (close to their melting temperature $T_m$), and we must control temperature accurately. This is the case for semi-crystalline plastics, which are difficult to thermoform and demand considerable skill and experience—examples are HDPE, PP, and PET (the latter is, in fact, a special case, but discussing it takes us beyond the scope of this chapter).

---

**EXERCISE 9.2**

Strictly seen from the manufacturing point of view, thermoforming is best done with high molecular weight grades. Explain why. Do you expect that such grades are also well-suited to injection molding? Explain your reasoning.

---

**EXERCISE 9.3**

Sketch the graph of stiffness versus temperature for both an amorphous and a semi-crystalline plastic. Indicate clearly the temperature ranges where you can thermoform these materials and where you can injection mold them. What range is suitable for the actual use of the materials in products?

---

Important as it is, the suitability of a certain plastic for thermoforming is not only determined by this graph. Thermoforming is an open process, offering only limited capability to control the atmosphere under which it takes place. Consequently, plastics that readily absorb moisture from the air when heated generally cannot be thermoformed well. This is the case, for instance, for polyamides ("nylons"). So the same property that makes these plastics comfortable to wear (as fabrics) makes it unsuited for thermoforming. It can be done, but only at relatively high cost.

---

## 9.3 THE BASICS: CONSERVATION OF VOLUME

The first basic rule that applies to thermoforming is that the available material volume is constant throughout the process. What goes in also comes out again, only in a different form. With this simple fact we can easily calculate the average wall thickness of the product. To do so, we make the following two assumptions:

1. There is a clear boundary between the cold material (which is not deformed) and the warm material (which is deformed).
2. The wall thickness of the product is constant everywhere.

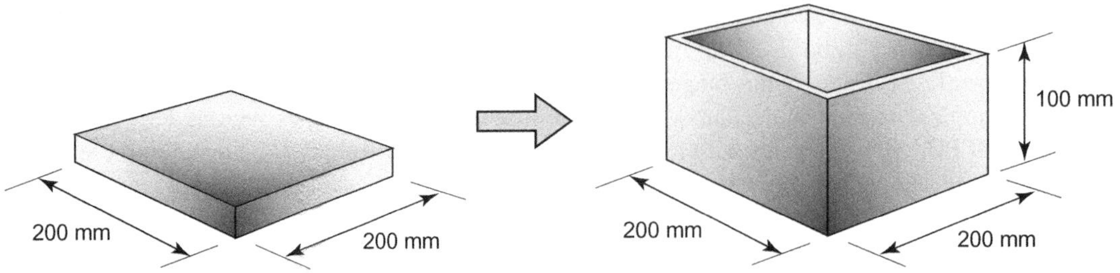

**FIGURE 9.1**

Thermoforming of a packaging box from a square sheet.

---

**EXERCISE 9.4**

We want to make a square PS box as shown in Figure 9.1. The semi-finished product we use as input material is a 1 mm thick sheet of PS. What will be the wall thickness of the box if it measures $L \times W \times H = 200 \times 200 \times 100$ mm? And what if we use a foil that is 0.2 mm thick?

---

**EXERCISE 9.5**

Suppose we want to thermoform a transparent hemispherical cupola from PC with a 3 mm wall thickness. What should be the thickness of the input sheet? Hint: the surface area of a sphere with radius $R$ is equal to $4\pi R^2$.

---

In practice, the first assumption is certainly reasonable, given that plastics have very low heat conductivity. The actual forming step in thermoforming generally takes much less time than it takes for the heat to diffuse from the heated area to the cold edge around it. The second assumption, however, is not so reasonable: in fact, a constant wall thickness is almost impossible to realize, especially for products with sharp radii. Thermoformed products will always be thinner at corners. So the "law of conservation of volume" only gives an *average* wall thickness.

---

**EXERCISE 9.6**

Why is the wall thickness at the corners less than in the middle? Hint: once cooled below $T_g$ the material will no longer deform, and further deformation has to come from the parts that are still hot.

---

**EXERCISE 9.7**

Take a close look at your sample products from Exercise 9.1. How does the thickness vary over the cross section? Tip: if appropriate, feel free to cut up your products, and use calipers to make accurate measurements.

---

**EXERCISE 9.8**

We can sometimes reduce this corner thinning effect by reducing the temperature difference between (heated) plastic and the mold. Does this also have a disadvantage? Explain.

---

## 9.4  A CLOSER LOOK AT THE PRINCIPLE

Thermoforming always follows five distinct steps, discussed later: (1) heating the material, (2) forming, (3) cooling, (4) de-molding, and (5) edge trimming. Steps 1 and 3 generally take up 80% to 90% of the cycle time, especially when the input material is thin, but all five are needed to get a product. As we will see, the principle allows for two basic choices: positive versus negative forming (in other words, forming *over* or *into* a mold), and forming with pressure or partial vacuum—the latter is often called "vacuum forming". Yet another distinction is between the forming of thin sheets and foils, also known as "thin-gauge thermoforming," and thick sheets, referred to as "heavy-gauge thermoforming." Anything below 2.5 mm counts as "thin," anything above that (fairly arbitrary) value counts as "thick."

Step 1 is the heating of the plastic sheet or foil. This can be done in different ways, depending, among other things, on the production volume and the part thickness. For large volumes, such as disposable coffee cups, the cycle time must be very short, and this means that only thin sheets or foils can be used, usually with infrared heaters. Foils are also stretched during heating, usually in one direction but sometimes also bi-axially. For smaller volumes, it is possible to use thicker materials (which take much more time to heat up and cool down). Then, ceramic heaters are generally preferred. If these are electrical heaters, as is the norm in the European Union, it is common to apply differential heating. Some areas of the sheet are heated more than others—for example, areas where relatively more deformation is needed. If gas burners are used, as is customary in the United States and other areas where gas is relatively cheap, this is not an option.

---

**EXERCISE 9.9**

What is the maximum cycle time if we want to make a million products per year with a single-cavity mold? Assume production in two shifts. Do you think this will be a product with large or small wall thicknesses? What advantage would a multi-cavity mold offer?

---

Most manufacturing processes discussed in this book involve two-shift production, with about 16 effective production hours per day (80 per week). This allows a decent return on investment for the production capital (in this case, thermoforming machines, plus the factory that contains them) without the high additional labor costs of continuous 24/7 production with 168 hours per week. Two-shift production is flexible: if demand drops, the company can temporarily switch to single-shift production, whereas bursts in output are obtained by having the employees work overtime.

Directly after heating comes step 2: the actual forming. The hot sheet or foil is clamped on all sides onto an airtight box, then it is formed using a pressure difference between the two sides. There are two

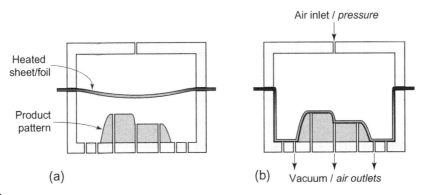

**FIGURE 9.2**

Schematic of vacuum/air pressure thermoforming: (a) heating stage; (b) forming stage. (Note: italic annotation refers to air pressure thermoforming.) (For color version of this figure, the reader is referred to the online version of this chapter.)

options: (1) using under-pressure, or partial vacuum (i.e., "vacuum forming") or (2) using over-pressure ("pressure forming"). In thin-gauge pressure forming, 6 bar is normally applied; in heavy-gauge thermoforming products, usually 3 bar is applied. It may be counter-intuitive that thinner products get more pressure than thicker ones, but as we will show in a moment, this can be explained if we look closer at what happens during forming. Figure 9.2 shows vacuum and pressure forming. Note that in both cases, we must expel the air beneath the product. This is done by means of an array of tiny holes (0.2 mm diameter) in the underside of the box. If you take a close look at thermoformed products, you can sometimes see the marks that these tiny air outlets leave on the surface.

---

**EXERCISE 9.10**

Take a good look at your sample products (Exercise 9.1). Any sign of air outlets? And do you think these products were vacuum formed or pressure formed? How can you tell?

---

**EXERCISE 9.11**

Think of one potential benefit of using higher pressure differences and one potential drawback. Explain.

---

The actual forming step can take place at high speed and sometimes lasts just a fraction of a second, but there is an upper limit to this speed. You can get a feel for this limit if you stretch chewing gum toward its yield point: if you do this slowly, you can stretch it much farther than if you try to pull it apart quickly. Thermoplastics show the same effect, known (somewhat ambiguously) in industry as "yielding." For heavy-gauge products, you have to slow down the forming speed to give the material sufficient time—seconds, not minutes—to form. With thin-gauge products, this is much less of a problem and forming can be done much faster. Note that in general, semi-crystalline plastics and materials with a small forming window (i.e., with a short or sloping rubber phase) are more sensitive to yielding than others, and this explains once more why PS, PVC, and ABS are more popular for thermoformed products than PP and HDPE.

When it comes to pressure forming, a word of caution: pressure differences not only act on the starting materials, but also on the box and on the molds. The higher the pressure, the stiffer we must make these parts to prevent excessive deformation. So there is a downside to using higher pressures. Of course, we would like to know in advance how much pressure is needed to create a certain shape, so we can decide between using vacuum or pressure forming. In "standard thermoforming" (see Section 9.6) we can increasingly rely on computer simulations to make such predictions, but for the other methods such tools are still a work in progress.

Using vacuum or pressure is not the only choice: we must also decide if we want to use *positive* forming (i.e., forming *over* a mold) or *negative* forming (i.e., forming *into* a mold). Figure 9.2 shows the first option. This is an important choice to make, as it strongly affects the thickness distribution over the product. Positively formed products tend to be thinnest at the edge, whereas negatively formed products are precisely at their thickest here. This is because areas of material that are cooled down through contact with the cold mold do not deform anymore, so that any further deformation has to come from areas that are still warm. Furthermore, notice that any texture, embossed symbols, and the like that are put on the mold only get transferred to the side of the product that actually makes contact with that mold. In Figure 9.2, this would be the underside of the sheet. There is, in fact, a third important consequence to this choice. When we want to use mult-cavity molds, then negative forming is effectively the only real option of the two, as positive forming would then cause undue differences in precise geometry and shrinkage between the various products. And when discussing de-molding (step 4) we will learn there is even a fourth consequence!

---

**EXERCISE 9.12**

A manufacturer of roof-mounted luggage compartments for cars wants to use thermoforming to shape the upper and lower shells of these products. The outside should be textured. Should the two shells be positively or negatively formed?

---

**EXERCISE 9.13**

Typical "clamshell packaging" only works if the edge is well-defined. Should we use positive or negative forming? And what about your sample products (Exercise 9.1)?

---

Step 3 is the cooling stage. The heated plastic is pressed against the cold mold by the pressure difference and cools down to a temperature below its $T_g$. How long this takes depends, among other things, on the (local) product thickness, the type of plastic we wish to form, and the mold material. For instance, if we use wooden molds, which are attractive for prototyping, then cooling takes much longer than if we use the aluminum molds (with or without built-in water cooling) that are typically used for high-volume production.

---

**EXERCISE 9.14**

State three additional factors that determine the cooling time, and explain if these factors increase or decrease the time.

---

De-molding, or taking the product out of the mold, is step 4. It is subtly different from de-molding in injection molding (or even in sheet metal forming, where we also remove a thin product out of the "mold," i.e., the die). Thermoformed products are often quite thin and flexible and can sometimes be forced from molds with negative draft angles (Figure 9.3). Of course, with heavy gauge thermoforming, such as with the inner panels of a common household fridge, this option will not be available, and de-molding requires us to incorporate positive draft angles into our part and mold design.

Note that the plastic will always shrink during the transition from warm to cold, especially if the plastic is semi-crystalline (e.g., HDPE, PP; see also Table 9.1). This may affect de-molding, depending on whether we use positive or negative thermoforming, as can the stiffness of the plastic (especially for polycarbonate, PC). But note that because the forces are relatively low, it is comparatively easy to use small sliding or rotating parts in the mold to make larger undercuts.

---

**EXERCISE 9.15**

What is easier: de-molding of positively formed products or de-molding of negatively formed ones? Explain.

---

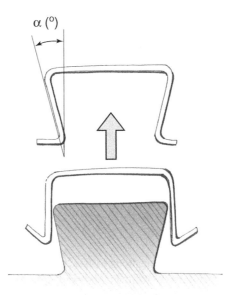

**FIGURE 9.3**

De-molding with negative draft angles. (For color version of this figure, the reader is referred to the online version of this chapter.)

**Table 9.1** Indicative thermoforming process data for several common thermoforming plastics (only applicable to high-volume production, simple shapes, low surface quality requirements, pressure forming)

| Plastic Type | Thermoforming Temperature Range | Typical Mold Temperature | Processing Shrinkage | Cycles for 1-mm Thickness |
|---|---|---|---|---|
| PS | 165–190°C | 80°C | 0.5% | 420/h |
| PVC | 155–200°C | 80°C | 0.4%-0.5% | 420/h |
| ABS | 160–220°C | 85°C | 0.6%-0.7% | 420/h |
| HDPE | 170–200°C | 100°C | 1.2%-7% | N/A* |
| PP | 160–200°C | 90°C | 1.5%-1.8% | 325/h |
| PC | 180–200°C | 130°C | 0.9%-1.1% | 300/h |

*Not applicable. This plastic type is only rarely used for thin-gauge parts.*

Finally, there is step 5: edge trimming (i.e., removal of the part of the input material that does not end up in the final product). Again, there is an important difference between relatively thin products and thicker ones. With thin-gauge thermoforming, edge trimming is almost exclusively done by cutting (and for larger production volumes, this step is usually integrated into the full process); with heavy-gauge thermoforming, it is done by machining (e.g., using a saw or a CNC-operated router). It can be done robotically, which costs more money but gives the added benefit that the robot can also perform additional operations, such as machining ventilation slits or stacking the finished products into crates for transport. Edge trimmings can be collected and recycled as production scrap, though this is not as easy as in injection molding, as the scrap must be extruded into foil or sheet again, which cannot be done in-house and therefore involves considerable transport.

**EXERCISE 9.16**

Take a close look at the edges of your sample products (Exercise 9.1). What do you find? More specifically, does this kind of edge finishing look and feel attractive?

## 9.5 DIGGING DEEPER: HEATING AND COOLING

Heating and cooling were discussed in the previous section, but there is more to say about these two steps. First, recall that it is difficult to rapidly heat up plastics due to their high heat capacity and poor heat conductivity. One solution that works well, at least for thin-gauge sheets, is to use electrically powered infrared heaters, as noted earlier. The ceramic heaters used for heavy-gauge sheets respond less rapidly than infrared heaters but are robust and reliable. In either case, heating can be done from one side, but for high-volume production, it is more common to heat from both sides. This *radiative* heating can be augmented by *convection* (i.e., by blowing hot air against the foil or sheet).

---

**EXERCISE 9.17**

Assuming that plastics are as sensitive to infrared as to normal visible light, what is faster to heat up: white sheets, black sheets, or transparent sheets?

---

Second, the product thickness $t$ has a strong influence on the heating time. Theoretically, if we double the thickness and keep everything else constant, we multiply the heating time by a factor of four: so the relationship between $t$ and heating time is quadratic, at least in the simplest of models. In practice, for a wide variety of reasons, the relationship is more complex, but the trend is the same: thicker parts take longer to heat. Then there are the large differences between the various types of plastics used in thermoforming. Generally speaking, semi-crystalline plastics require somewhat more energy to heat up than amorphous ones because the melting enthalpy of the crystalline parts now comes in also. Apart from that factor, the temperatures at which they can be formed are, of course, different. Finally, the exact geometry and surface detail that are required have an effect on the heating time. Deep shapes requiring large deformations and shapes with sharp details and radii require processing temperatures at the high end of what is given in Table 9.1.

---

**EXERCISE 9.18**

Look at your sample products (Exercise 9.1) and try to sort them based on heating time. Which one has probably taken the longest time to heat, which one the shortest? Why?

---

Now, let us look closer at cooling. The color or transparency of the material obviously does not affect its cooling time, but its thickness does. The relationship between time and thickness is complicated by the fact that there is only good heat transfer on the one side in contact with the mold. When semi-crystalline plastics are involved, thickness is even more influential, as the thermal gradient between the side that makes contact with the mold and the other side can easily cause undue distortion: the contact side crystallizes and shrinks considerably, whereas the other side is not yet crystallized. Increasing the mold temperature can mitigate this effect, but cooling will take longer. The same solution (and slower cooling) is often needed when we have extreme forms or extra demands on the thickness distribution over the part cross-section. Cooling itself can be accelerated with several "tricks," such as using fans to blow air over the part (i.e., forced convection, again) or spraying a water mist onto it. Finally, note that high forming pressure will reduce cooling time somewhat, as it improves the thermal contact between the part and the mold.

---

**EXERCISE 9.19**

Make a schematic graph of part temperature against time during cooling, showing the forming temperature $T_f$, de-moulding temperature $T_e$ and mould temperature $T_d$ (with $T_f > T_e > T_d$). If the cooling time is the time to get from $T_f$ to $T_e$, when is this time then minimal? For an amorphous plastic, how close to $T_f$ can we de-mold?

---

---

**EXERCISE 9.20**

Which could be faster for cooling: positive or negative forming? Hint: think of shrinkage.

---

**EXERCISE 9.21**

Revisit Exercises 9.8, 9.11, and 9.14. Now, sort all factors that influence the heating and cooling time into principle-specific factors and method/equipment-specific factors. Explain.

---

With so many factors at work, there is clearly no meaningful, representative cycle time for any given part. Table 9.1 just gives a first, rough indication of the cycle time for a 1 mm thick part in the various plastics, assuming high-volume pressure forming of simple shape, with low requirements for surface quality and finish, with two-sided heating and forced convection.

Finally, a word of caution: the lower the processing temperature we choose, the faster that the cooling phase will be, all else being equal. So from the perspective of productivity (i.e., cost), low forming temperatures are good. But as so often is the case, this is not good for quality, as we then get more built-in stresses and a lower resistance of the part to high temperatures (recall that plastics are viscoelastic and will try to revert to their original shape, in this case a sheet or foil, when exposed to heat) or even just to sunlight. A simple, visual inspection of a formed part will not reveal if this problem is present, so the supplier and designer need to agree how to handle these issues for more demanding applications.

---

## 9.6 THERMOFORMING METHODS

We can sort the various methods of thermoforming into categories depending on the number of forming steps and the number of product-specific investments that are needed. Each method can be economic over a range of production volumes—sometimes a very wide range—and when high volumes are involved, additional investments are usually needed, such as dedicated facilities for feeding and pre-heating of input material, automated stacking and packaging equipment, and so forth. This diversity can be seen in nearly every manufacturing principle, but nowhere is it as wide as in thermoforming.

### Free-blowing

This method has minimal product-specific investments. A sheet of plastic is clamped, heated, and simply blown upward into the desired shape (usually a dome of some sort) and then cooled in air. Because there is no cold mold, cooling takes comparatively long, with cycle times in minutes rather than seconds. Form freedom is very limited, but it is sufficient for products like skylight cupolas.

## Standard thermoforming

This is the method implicitly referred to in the previous sections, with the four combinations of pressure or vacuum and positive or negative molds. What these four have in common is that they all use one product-specific investment, the mold, and forming is done in one step. Another common factor is that part wall thickness, and its distribution over the cross-section, is difficult to control. Furthermore, parts should not be too deep and too complex—other methods are better suited for such shapes. There is considerable design freedom when it comes to materials and product size, which can range from tiny (e.g., 10 cm$^3$ packaging cups) to huge (e.g., the front end of a caravan). Production volumes also vary widely (perhaps more than for any other manufacturing method, for any principle): from "one-off" prototypes using wooden molds to hundreds of millions per year (using complex, 48-cavity molds and a fully dedicated production system). As another example, standard 250 cm$^3$ PP drinking cups are produced at more than 75,000 parts per hour, using 32-cavity molds and 40 cycles per minute. Of course, this is a thin-gauge product, but even for 4 mm thick ABS inner fridge panels, cycle times of just 40 seconds are possible, allowing large production volumes with notably lower investments than for injection molding.

## Pre-stretch thermoforming

This method generates a more homogeneous thickness distribution; it is also well-suited for more extreme forms (deep and complex shapes with sharp surface detail). It works by first blowing the sheet or foil away from the mold, pre-stretching it, then making the actual forming step. Figure 9.4 shows the two steps in succession. Like standard thermoforming, there are several variants (some with catchy names such as "vacuum snap-back forming"). The downside to the form freedom of this method is somewhat extra complexity, slightly higher cycle times and, hence, increased cost.

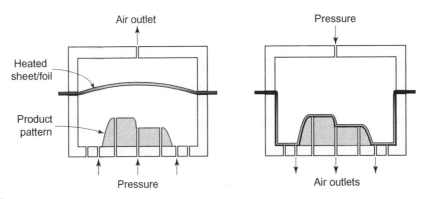

**FIGURE 9.4**

Schematic of pre-stretch thermoforming. (For color version of this figure, the reader is referred to the online version of this chapter.)

## Plug-assisted thermoforming

If, while pre-stretching the sheet or foil, we blow it against a cold "plug" before the actual forming step, we can create a part that is thicker in one area (i.e., the area that made contact with the plug) than elsewhere in the part. This method is shown schematically in Figure 9.5. So we now have two forming steps

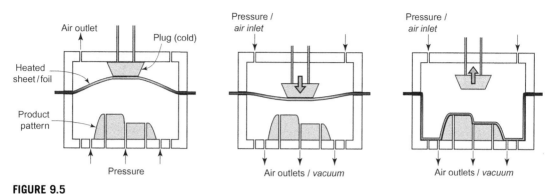

**FIGURE 9.5**

Schematic of plug-assisted thermoforming. (For color version of this figure, the reader is referred to the online version of this chapter.)

and two part-specific investments: the mold and the plug. It is well-suited to manufacturing products where, for example, part of the shape must have a closely toleranced thickness.

## Twin-sheet thermoforming

In this recently developed method, we heat and form two sheets, each with its own mold, and press the two together before the material is fully cooled down. This allows us to create hollow parts, such as containers for liquids or plastic freight pallets. Such products are lightweight, yet strong and stiff and often surprisingly rugged. The method competes with injection blow molding or extrusion blow molding, generally winning when products are relatively large.

## Cost breakdown: some additional comments

In thermoforming, the materials tend to be responsible for a considerable part of the total production cost: typically 30% to 50% for thin-gauge products and 45% to 70% for heavy-gauge ones. In economically developed countries, the costs of operating the forming machines themselves range between 100 to 300 euro/hour. So designing your parts to be as thin as possible gives a double benefit: lower material costs and, through faster cycle times, lower machine costs. Transport of thermoformed products is relatively expensive, as the products tend to be light or even contain a lot of air. For this reason, thermoforming companies limit their market to clients within a 750-km radius around their location. For stretch blow-molded bottles, the preform parisons are relatively compact, so it pays in transport terms to blow the bottles to full size in the same facility as they will be filled.

Note that when production volumes get really large, as they are for coffee cups, the extruder that produces the sheet or foil is usually placed in line with the thermoforming machine, together forming a single production line. This has two major cost benefits: (1) it saves energy, as the plastic only needs to be heated once, and (2) it allows easy in-house recycling of production scrap (the material that is trimmed off can simply be reground and put back into the extruder, without the potentially costly transport step that is involved if the extruder and thermoforming machine are in different locations).

---

**EXERCISE 9.22**

Sort the aforementioned five methods by (a) form freedom (think of product size and depth, level of detail, thickness control) and (b) production speed. What is your conclusion?

---

**EXERCISE 9.23**

Write down three guidelines for the design of successful thermoforming products, with reference to the theory described in this chapter. How can you see these rules reflected in your sample products (see Exercise 9.1)?

---

## 9.7 Further Reading

As always, this chapter has limited itself to the main points, leaving out many details, particularly on quality issues. Similarly, it is worth researching into special thermoforming materials, such as C-PET, a highly crystalline, heat-resistant grade of PET that has revolutionized convenience food packaging since the 1980s. More recent developments to investigate concern PA (nylon) and POM, two plastic types that have only been thermoformed since the early 2000s.

One additional method of particular interest is *press forming*, in which pressure is applied onto the heated sheet directly by matching metal molds instead of by air pressure (so it is similar to matched die forming of metal parts). When applied to advanced thermoplastic fiber composites, this method has great potential—for example, these novel materials currently see application in the leading edges of the Airbus A380 superjumbo's wings, where low weight and high impact strength are crucial. The method has been in use since the 1980s to form conventional glass mat reinforced thermoplastics to produce all kinds of semi-structural automotive components. The BMW Z1 roadster with its unique roll-down doors was a notable "early adopter," but higher volume applications exist as well.

The website www.t-form.eu, an output of a European research project into the method, is a good starting point to explore the thermoforming process. For academic and industrial journals and trade fairs, refer to the previous chapter on injection molding. Most books on plastic processing will contain a section on thermoforming, or at least on "vacuum forming" (one of its variants). Sources that specifically address thermoforming are as follows:

Throne, J.L., 1996. Technology of Thermoforming. Hanser Publishers, New York. (*Generally considered the number one reference for thermoformers worldwide, this seminal work describes the full breadth and depth of the process. Note this quote: "Thermoforming is not an easy process. It just looks easy".*)

European Thermoforming Division, An Easy Guide to Heavy Gauge Thermoforming Technology www.e-t-d.org. (*More accessible than the previous source and good additional reading.*)

# Resin Transfer Molding

# 10

## CHAPTER OUTLINE

With its resin transfer molded, carbon fiber composite body, the BMW i3 electric city car represents a breakthrough in automotive manufacturing engineering. At a cycle time of only five minutes, its body is fully suitable for mass manufacture.

*Image courtesy BMW.*

## 10.1  INTRODUCTION

Few materials make such a strong appeal to the imagination as fiber-reinforced plastics, also known as *fiber composites*. They have a reputation for being very light, stiff, and strong, and have a modern, 'high-tech' image associated with exciting applications, from sports products and Formula One cars to military jets. But fiber composites have been around for longer than you may think. For instance, glass fiber composites have been used since the 1950s, those with the stiffer (and more expensive) carbon fibers since the late 1960s, and composites based on the tough aramid fibers, known under the trade names of *Kevlar* and *Twaron*, since the 1970s. And humankind has used straw-filled clay to build huts for thousands of years!

Regarding manufacture, composites are special because the material is often created not before, but during the manufacture of a part. This is the case in the *resin transfer molding* (*RTM*) process, in which fibers are first placed in a closed mold, followed by injection of a thermosetting plastic resin into the mold, making a continuous matrix around the fibers. RTM was first deployed commercially in the 1970s and is now well developed. It is typically well-suited to production volumes up to 10,000 units/year, offering considerable form freedom and good potential for functional integration. As an example, the bodywork of the famous Lotus Esprit consisted of just two parts: the upper half and the lower half (made with RTM since 1987). The more recent BMW i3 also uses a variant of this process and represents a breakthrough in high-volume mass production of carbon fiber composites.

RTM is by no means the only process for making composite parts and products. For instance, there are *hand lay-up production* (suitable for prototypes and very small production volumes), *filament winding* (suitable for large, rotationally symmetrical shapes), *press-forming* of thermoplastic composites (comparable to thermoforming; see Chapter 9), *pultrusion*, and *autoclaving*. These processes generally use some type of *pre-preg* (i.e., uncured resin "pre-impregnated" with fibers). In this chapter, we start with some general observations on composites applicable to all these processes, but we then focus on RTM only. As usual, RTM can be done using several methods, so we will work down from general principles to explain the physical basis of the process. For other composite processes, refer to the "Further Reading" section.

---

**EXERCISE 10.1**

Search the web for interesting examples of RTM products. Even better, find actual products, if you can get your hands on them! For each product, why do you think a composite has been selected rather than an unreinforced plastic? What are the implications of this choice for recycling?

---

## 10.2  FIBERS, RESINS, AND COMPOSITES: AN INTRODUCTION

To understand the nature of RTM, we first explore the various fibers and resins that are used today. The properties of composites depend on (1) the properties of the fibers and matrix resin, (2) the volume fraction of fiber, and (3) the fiber orientations. Note that the focus is on composites with *continuous* fibers, not on those with chopped pieces of fiber—the differences in strength and stiffness are huge. Table 10.1 lists the main properties of several common fibers and resins, as well as representative data

**Table 10.1** Relevant Properties of Various Fibers and Resins

| Material | Tensile Strength [MPa] | E-Modulus [GPa] | Density [$\times 10^3$ kg/m$^3$] | Elongation [%] | Price (2013) [€/kg] |
|---|---|---|---|---|---|
| E-glass fiber | 1,950-2,050 | 72-85 | 2.55-2.60 | 2-3 | 1.45-2.89 |
| R/S-glass fiber | 4,700-4,800 | 86-93 | 2.49-2.50 | 5 | 17-19 |
| Aramid fiber (K29)* | 2,900-3,600 | 62-80 | 1.43-1.45 | 2.5-4.4 | 22-36 |
| Carbon fiber (HM)* | 2,400-2,410 | 370-640 | 1.80-2.10 | < 2 | 35-42 |
| Carbon fiber (HT)* | 4,500-4,800 | 225-245 | 1.80-1.84 | < 2 | 18-30 |
| Epoxy resin | 70 | 3.2 | 1.00-1.15 | 5-7 | 4-10 |
| Polyester resin | 50 | 1.5 | 1.00-1.10 | 2-4 | 1-4 |
| GFRP[†] (typical) | 500 | 20 | 1.80 | < 2 | 19-27 |
| KFRP[†] (typical) | 800 | 23 | 1.25 | < 2 | 38-63 |
| CFRP[†] (typical) | 1,200 | 56 | 1.40 | < 2 | 30-33 |
| HSLA steel[‡] | 400-750 | 205-210 | 7.80-7.90 | 10-30 | 0.30-0.50 |
| Aluminum 6061-T6 | 241-320 | 68-74 | 2.67-2.73 | 5-12 | 1.90-2.10 |

*K29, Kevlar type 29; HM, high modulus; HT, high tenacity.
[†]E-glass-, HT-carbon-, and aramid (Kevlar)-fiber reinforced plastic, respectively. Approximate data for typical epoxy-based composites with 50% fiber volume (0/90), cross-ply fiber lay-up.
[‡]High Strength Low Alloy steel (currently very popular in the automotive industry).
Source: Cambridge Engineering Selector (CES), Granta Design, Cambridge, UK.

for the composites made from them. In RTM, the resin is by definition a thermosetting plastic resin that enters the process as a viscous liquid and "cures" (i.e., sets) during the process. Most other composite manufacturing processes used today involve thermoset resins (in the form of pre-pregs), with only a few using thermoplastics. The resin properties in Table 10.1 apply to the as-cured condition. For comparison, Table 10.1 also lists data for structural steels and an aluminum alloy that offer good strength at acceptable cost. Note that all fibers are inherently brittle, showing linear elastic behavior until fracture—unlike metals they do not have a yield stress and show no plasticity. But their strengths are nonetheless very high.

**EXERCISE 10.2**

Compare the specific stiffness (modulus/density) and specific strength (strength/density) of the five fibers, HSLA steel, and aluminum. Which one "wins," and how do the metals come across in the comparison?

It will be clear that especially the more expensive fibers are light, stiff, and strong. However, the fibers cannot function if there is no resin to keep them in place, allowing for handling and carrying structural loads (particularly bending and compression—ropes, for example, are loaded in tension and consist purely of twisted fibers). In high performance composites, the fibers take up around 50% to 60% of the total volume; for less demanding applications this fiber volume fraction can fall to 20%. To a first-order approximation, we may assume that, in an aligned "unidirectional" (UD) composite, the fibers carry the entire load (for in-plane loads only)—this is a reasonable starting point when the fiber stiffness and strength are much greater than those of the resin, as is clearly the case in Table 10.1. The density, however, is given by a simple linear rule of mixtures. So the necessary addition of resin effectively has a strong "knockdown" effect on the mechanical properties, relative to the weight.

---

**EXERCISE 10.3**

Choose a combination of fiber and resin. Estimate the strength and stiffness of a unidirectional composite that has a fiber volume percentage of 50% (on the basis that the fibers carry all of the load). Now repeat for a value of 20%.

---

**EXERCISE 10.4**

Increasing the fiber volume percentage clearly benefits strength and stiffness, but what could be the benefit(s) of lowering this percentage?

---

A second important factor in the actual strength and stiffness of composites is the fiber orientation. The mechanical properties of the fibers listed in Table 10.1 are only fully exploited when they are all aligned parallel to the load. But perpendicular to the fibers, the weak, flexible resin carries much of the load. Even if the external load comes from one direction, it is rare to use a unidirectional (UD) fiber composite. This is because small amounts of bending or any off-axis load tends to cause *splitting* parallel to the fibers (an issue with cheap composite tent poles). In practical composites, the fibers are therefore oriented in several directions—Figure 10.1 shows various options. The simplest is a biaxial arrangement of UD layers, giving a "0/90 cross-ply" composite laminate. A further option is to extend this approach to include ±45-degree orientations, giving "quasi-isotropic" composite laminates. Another possibility is to make a composite that really is isotropic in the plane of sheet, by giving the fibers a random orientation. In such a "random mat" composite, the fibers are not straight (as in a UD or laminated composite).

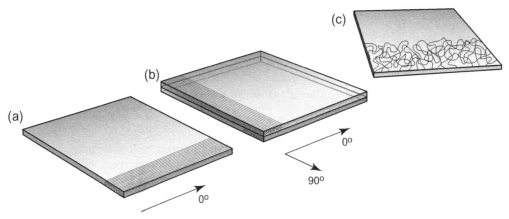

**FIGURE 10.1**

Long fiber composites: (a) unidirectional (UD), (b) cross-ply (0/90), and (c) random mat. (For color version of this figure, the reader is referred to the online version of this chapter.)

---

**EXERCISE 10.5**

Repeat Exercise 10.3, but this time for a cross-ply composite with equal amounts of fibers in both directions, and neglecting the stiffness and strength of the transverse (90) plies (total fiber volume: 50%). Now estimate the specific stiffness and specific strength of your composite, and compare again with the metals in Table 10.1. What are your conclusions?

---

**EXERCISE 10.6**

Even with equal fiber volume percentages, a random mat composite has a lower strength and stiffness than a quasi-isotropic composite (at least for in-plane loads). Explain why.

---

**EXERCISE 10.7**

Fiber layers always have a certain thickness, which is typically around 0.1 mm. What does this mean for the available thicknesses of UD, cross-ply, and quasi-isotropic composites? Note that fiber layers should preferably be stacked symmetrically over the thickness!

---

The high strength of the fibers allows composites to compete easily with metals, even with the two knockdown effects of fiber volume and fiber orientation. In fact, these effects unlock a particular design freedom that metals do not offer: we can configure materials with exactly the right strength or stiffness, and we can even create controlled anisotropy, producing materials that match the loads they must carry exactly. Combining different kinds of fibers further extends our range for optimization: for instance, we can combine carbon and glass fibers to "tune" the price-performance ratio, or add aramid layers to a carbon-fiber core to increase impact strength. There are even specialist laminates, in which panels are made by stacking thin layers of GFRP and aluminum alloys (known as 'GLARE'). These laminates are increasingly used in jet aircraft fuselages (e.g., in the Airbus A380).

A final point: another way to assist in the handling of fibers during manufacture is for them to be woven or stitched together, giving a mat or woven pre-preg. This merges these relatively new materials with the historic fields of weaving and textile technology—indeed the field of "technical textiles" embraces many advanced architectures of fiber-reinforced plastics, used extensively in civil engineering and defense. In woven composites we then have additional parameters associated with the weave, which not only influence the effective strength and stiffness but also the "drapability" (i.e., how easily the fabric can be bent, particularly with double-curvature). The weave also affects the look-and-feel: for instance, the tell-tale pattern of small squares seen on many carbon-fiber bicycle frames is a result of how the fibers are woven.

---

**EXERCISE 10.8**

Which fibers have been used in your sample products (Exercise 10.1)?

Composite technology has become a broad, mature field, though it continues to be dominated by high performance or high value-added structural applications (sports goods, aerospace, and defense). In line with the selective nature of this book, we now focus on the composite manufacturing process that gives the most form freedom at relatively low cost (i.e., RTM), as this competes most directly with the other plastic processing principles explored in Chapters 8 and 9.

## 10.3 MOLD FILLING DURING RTM: d'ARCY's EQUATION

Figure 10.2 shows in three steps how RTM is done, with an upper and a lower mold that jointly enclose the part. In step 1, the fibers (and any inserts, such as a lightweight foam core) are placed in the mold. Then, in step 2, the resin is transferred from an external container into the mold, driven by a pressure difference $\Delta p$. One common way to do this is to inject the resin into the mold from the outside inward by means of a "runner" that traces the entire contour of the part, with the resin flowing toward one or more central suction cups. Another way is to inject across the product width, from one side to the other. As soon as the mold is fully filled, step 3 commences: the "curing" (setting) of the resin, followed by de-molding. To give a first idea of the time scale involved, for a large suitcase, the total cycle time can easily be one hour.

Low-viscosity fluids, such as water or alcohol, would fill an empty mold easily and quickly under gravity. However, resins are considerably more viscous (water has a viscosity of 0.001 Pa.s, against typically 0.2 Pa.s for RTM resins). Moreover, the fibers placed inside the mold cause considerable resistance to flow. To achieve the filling speeds needed for high volume production, RTM always requires a pressure difference. The question is, how much?

The French scientist d'Arcy was the first to answer this important and interesting question. He found a simple formula for the speed $v$ of a liquid through a porous medium. This depends on the pressure difference $\Delta p$, the length of the flow $l$, the viscosity of the liquid $\mu$, and the medium's "permeability" $k$:

$$v = (\Delta p\, k)/(l\mu) \tag{10.1}$$

where permeability $k$ has units of $m^2$. The fiber mats used in RTM qualify as porous media, so we can use d'Arcy's equation to gain a first insight into the process. Note that the value of $k$ will fall (and with

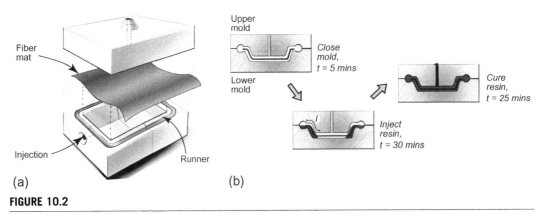

**(a)**

**(b)**

**FIGURE 10.2**

Schematic of RTM: (a) mold layout, and (b) process sequence, in cross-section.

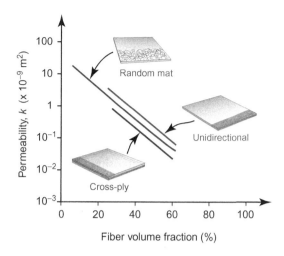

**FIGURE 10.3**

Relationship between fiber volume, fiber orientation, and permeability. (For color version of this figure, the reader is referred to the online version of this chapter.)

it, the injection speed $v$) if the fiber volume percentage increases (recall Exercise 10.4). Furthermore, $k$ depends on the direction relative to the fibers: parallel to the main fiber orientation it is higher than perpendicular to that orientation, and the flow speed will vary accordingly along the various directions. Figure 10.3 shows these relationships. Note the logarithmic scale: small changes in fiber volume strongly affect $k$.

---

**EXERCISE 10.9**

Assume we use RTM to produce each half of a rectangular suitcase measuring $90 \times 60 \times 20$ cm (so the depth of each half is 10 cm). Injection is done using a runner around the product and a single suction cup in the middle. Plot how the injection speed varies with time during the process, assuming that $\Delta p = 0.8$ bar, $\mu = 0.2$ Pa.s, and $k = 0.8 \times 10^{-9}$ m$^2$.

---

Reworking Equation (10.1) and assuming that the pressure difference $\Delta p$ is constant, we can obtain a second equation for the injection time $T_{inj}$ as given next. In this formula, $L_{inj}$ is the maximum value for length $l$, also referred to as the "injection length." In the situation shown in Figure 10.2, this is simply the shortest distance from the runner to the suction cup.

$$T_{inj} = L_{inj}^{2} \mu / (2 \, \Delta p \, k) \tag{10.2}$$

---

**EXERCISE 10.10**

What is the injection length for the situation in Exercise 10.9? Hence, determine the injection time. What would this time be if the injection length were doubled?

---

---

**EXERCISE 10.11**

By increasing the pressure difference, we can decrease the injection time, which is generally a good thing. However, does this approach have a drawback?

---

With the typical permeabilities, viscosities, and injection lengths involved in RTM, we would require hundreds of bars for injection times to get anywhere near to those in injection molding of thermoplastics. Although this is not technically impossible, it would require much stronger, more expensive, molds than is economical for the production volumes typical for RTM. Furthermore, the resulting high flow speeds would dislodge the relatively fragile fiber mats, especially at the injection points. In practice, therefore, RTM pressure differences are modest, around 1 to 6 bar, and injection times range from several minutes to around one hour (depending on product size and fiber volume fraction). Still, we should not underestimate the forces involved: for example, 1 bar on a product with a projected area of $1\,m^2$ still gives a force of 100 kN, or ten tons! It is not easy to make molds that can withstand such forces without significant deformation, especially when they have to be cheap as well.

---

**EXERCISE 10.12**

Calculate the total force acting on the suitcase featured in Exercise 10.9 during production (don't forget the runner!). Are the mold halves pressed toward or away from each other? Assuming that some mold deformation occurs, what happens to the permeability?

---

## 10.4 RESIN CURING: EPOXIES VERSUS POLYESTERS

After the mold has been filled, the resin needs to cure to a point where the product is strong enough to allow de-molding. For curing, the resin is a mixture with a second compound. In epoxy resins, this is done quite differently than in polyesters, with several key consequences. Note that curing causes a sharp increase in viscosity and should definitely *not* start before the mold is fully filled, or there will be a steep rise in the pressure required to complete filling.

*Epoxy resins* are mixed with a "hardener." This chemical substance (usually an amine compound) creates cross-links in the resin, causing it to set. The ratio between resin and hardener must be carefully controlled: too much hardener makes the epoxy very brittle, too little makes it too flexible. For details on resin chemistry, refer to specialist literature, but you should at least know that epoxies can easily take one hour to cure. The reaction is also exothermic (i.e., it releases heat), but because the reaction time is so long and the parts are relatively thin and between metal molds, dissipating the heat is not normally a problem. Note that all hardeners are by definition aggressive chemicals that must be treated with care to prevent damage to people or the environment.

**EXERCISE 10.13**

Assuming we use RTM to produce a carbon fiber-epoxy bicycle frame, roughly how many products can we make per year if we have one mold?

**EXERCISE 10.14**

If injection and curing both take one hour, what would be the benefit of having more than one mold?

*Polyester resins* are mixed with a "starter" compound (usually hydrogen peroxide), plus a small amount of cobalt that acts as a catalyst. The more starter we add, the faster the reaction takes place. This is very useful, because by gradually increasing the amount of starter during injection, we can obtain the ideal situation where the product begins to set homogeneously just when the injection has been completed (this requires careful control over the injection time, which favors certain RTM methods, as discussed later). Moreover, polyesters set much faster than epoxies, sometimes as quickly as 10 minutes. This may cause problems with temperature build-up, as polyesters set exothermically as well. In fact, the rising temperature speeds up the chemical reaction, risking a "runaway" situation, and deterioration of material properties. Additionally, many polyester-based composite parts have a high resin volume percentage, which exacerbates the heat dissipation issue. However, polyester resins can be used for production volumes up to 10,000 units per mold per year, depending on part size.

**EXERCISE 10.15**

Epoxies and polyesters both shrink some 4% during curing. For parts with a low fiber volume percentage, this shrinkage may lead to tolerance issues. Explain why this is generally less of a problem for parts with a high fiber volume percentage.

Not only do polyesters cure considerably faster than epoxies, but they are also available in many more variants, allowing more process optimization. So from an RTM manufacturing perspective, polyesters can be favored over epoxies. On the other hand, epoxies are tougher than polyesters and better suited for high performance products, such as carbon composite parts. The trade-off between function and cost should be obvious!

The resin industry has in fact developed a compromise between epoxies and polyesters in the form of the *vinyl esters*, which are slowly gaining popularity. Another development in this field, now finding its first applications, concerns bio-based resins (i.e., resins that are produced on the basis of renewable resources such as biomass instead of fossil fuels). For large production volumes, it may be possible to have a customized resin developed specifically for your application, as happened for the BMW i3 car (which, to make a noteworthy addition, is made using very high injection pressures and very strong steel molds).

**EXERCISE 10.16**

Which other manufacturing step(s) are needed to obtain a finished part, apart from fiber placement, injection, and curing?

As in bulk plastics, the resins used in RTM can contain additives, such as flame retardants or pigments. They can be opaque and colored as desired, or, alternatively, they can be fully transparent, showing off the fibers. For improved look-and-feel, it is common to use a gelcoat or clearcoat: these are thick resins that are brushed or sprayed onto the mold surface before the fiber mats are placed and that cure with the injected resin. Appearance can be spectacular, but at the cost of some extra weight and thickness (typically 0.5 mm).

## 10.5　DIGGING DEEPER: DESIGNING THE DETAILS

The formula developed by d'Arcy gives us a first, quantitative insight into the nature of RTM as a manufacturing principle. In this sense, it serves the same purpose as Chvorinov's rule for casting of metals. And just as we can apply that rule to understand both the full product as well as some of its details (e.g., how changes in section thickness can give rise to thermal stresses), we can apply d'Arcy's formula to understand both global and local problems. Two common local details encountered in RTM products are (1) variations in fiber volume fraction, and (2) internal runners.

Localized variations in fiber volume fraction cause variations in permeability and, as a result, variations in the local resin flow speed; see Equation (10.1). Usually the fibers stacked in the mold are an assembly of different precut patches. Where these patches overlap (intentionally or not), the fiber volume fraction increases, because at those locations, the mold is filled with more layers of fibers than elsewhere in the part. This makes it more difficult for the resin to flow through these areas (because of lower permeability; see Figure 10.4). When this effect is strong, it can even result in "dry spots"— that is, areas where the fibers are not (fully) impregnated by the resin. The part must then be rejected. So together, the designer and manufacturer have to deal with the potential negative effects on resin flow caused by possible overlaps of the different fiber patches.

---

**EXERCISE 10.17**

Suppose we make a product with four fiber layers and a 40% fiber volume fraction. How does the permeability change in areas where we add a fifth layer? Use Figure 10.3 to obtain a quantitative answer.

---

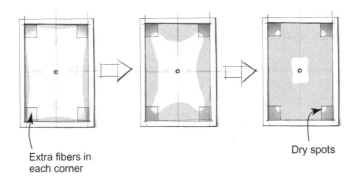

Extra fibers in each corner

Dry spots

**FIGURE 10.4**

Formation of dry spots.

Think of a solution to the problem raised in Exercise 10.17. Hint: what if we can somehow assure that the resin flow front always stays parallel to the area with decreased permeability?

In contrast, volumes with high permeability, referred to as "internal runners," have the opposite effect. Resistance to resin flow is low, and the result is locally higher flow speeds. Internal runners easily occur as a result of poor design, for instance, when we make the part's radii too sharp. In practice, it is nearly impossible to tuck fibers into sharp corners. Instead, fibers will "cut off" the radius, leaving an open space on the outside (Figure 10.5). The resin flow will take the path of least resistance to reach the suction cup, and these runners can result in products that are not fully impregnated on the inside of the radius and therefore must be rejected.

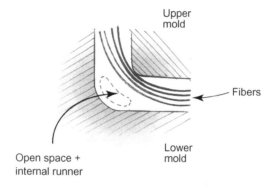

**FIGURE 10.5**

An internal runner caused by (too) sharp corners (cross-sectional view).

Using "internal runners" can also be deployed as a deliberate design strategy to minimize injection time. Can you describe how this works, in words or a sketch?

It would be nice if we could give a single lower limit for corner radii. However, this limit strongly depends on the kind of fiber mats we use. For instance, plain layers that are stitched together for handling can be tucked into corners somewhat better than woven mats can, as the stitching allows the layers to slide over one another. The fiber mat orientation matters as well: a 0/90 cross-ply will not deform much if we pull it parallel to the fibers, but if we rotate it by 45 degrees we find that the fiber architecture now allows deformation by shear—that is, "on the bias," something that clothing designers certainly know how to exploit! So by orienting the fibers at 45 degrees relative to the bend line, we can get considerably sharper radii than at 0 degrees. In general, the drapability of fiber mats is greatly improved if they are deformed in shear instead of in tension. But eventually the maximum strain that the fiber mat can withstand will set a lower limit to the bending radius.

---

**EXERCISE 10.20**

Take a closer look at your sample products (Exercise 10.1). How large are the radii that you can see?

---

## 10.6 MANUFACTURING METHODS FOR RTM

Because RTM is still a fairly young process, the various methods currently found in industrial practice cannot yet be categorized unequivocally. The fact that the same method can go by different names does little to help: for instance, RTM is sometimes also known as "(structural) resin injection molding," or (S)RIM for short. Still, we can impose a useful kind of order by looking at the number of part-specific tools: is there one rigid mold half or two? A second variable is the use of either pressure or vacuum to generate the pressure difference. (Recall that we used the same two variables to categorize the various manufacturing methods of thermoforming in Chapter 9.) A third variable is the manner of injection, which can be either pressure-controlled or volume-controlled. Four common RTM methods are presented next. As we shall see, the first and second variables are closely linked.

One important remark to start: from an environmental perspective, the closed-mold RTM process represents a significant step forward as compared to open processes, such as hand lay-up. However, the use of reactive chemicals inherently presents a certain risk to humans and the environment, and taking proper precautions is therefore always required.

### Pressure-controlled RTM

This RTM method involves two rigid mold halves, clamped together. The resin is injected at constant pressure (up to 6 bar is common, but in special cases this can be much higher) and flows rapidly at first, but gradually slower into the mold. If desired, a pressure profile can be employed, gradually ramping it up: this prevents the fiber mat from being deformed or even dislodged by the initial jet of resin that comes rushing in. The method relies on over-pressure and, hence, proper sealing of the mold halves requires attention. (Fortunately, resin viscosity is so high that the resin does not squirt out if there happens to be a small leak.) One disadvantage is that this method does not allow close control over the injection time and hence little optimization of the full production cycle. This gets especially problematic at higher fiber volume fractions, where the inevitable small variations in permeability cause large variations in injection time.

Typical products made with this method are flat, single-curved or (mildly) double-curved shell-shaped parts. Size can range from 0.1 to 6 m for the smallest and largest part dimension, depending on fiber volume fraction. Thickness is also constrained, with 1 mm being a practical minimum and few products thicker than 20 mm. For such thick composite parts we would usually make a "sandwich structure," consisting of a foam core and composite facings. As for form freedom, the first requirement is that the part should have generous radii, no undercuts, and a decent draft angle to enable de-molding. Inserts and foam cores of all kinds can be used and if these are firmly fixed, it is possible to achieve fairly close tolerances (but remember Exercises 10.14 and 10.15). Because the part is fully enclosed by the rigid mold, a good two-sided surface finish can be obtained and textured if desired.

Two notable applications of this method are car bumpers (done, for example, by Citroen, with numerous water-cooled steel molds in parallel and a cycle time of just 10 min/mold) and lightweight, easy-to-install roof dormers. Both are glass-polyester composite parts with modest fiber volume fractions.

## Volume-controlled RTM

This method is basically the same as the previous one, but now it is not the pressure that is kept constant, but the injection speed, and with it, the volume flow (expressed in cm$^3$/sec). Consequently, the pressure gradually rises as the injection progresses. This is more complex than pressure-controlled RTM, but because the injection time is now closely controlled, it allows for better optimization, even at higher fiber volume fractions (variations in permeability are now compensated for by varying pressure). Furthermore, it allows for easy automation: by putting a narrow opening at the end of the injection length, we can let the pressure rise steeply as soon as the resin arrives there. A simple pressure switch can automatically end the injection. Volume-controlled RTM is a robust, flexible process that is well-suited to relatively large production volumes; as for shapes and sizes, it is comparable to the previous method.

## Resin infusion (or vacuum bagging, vacuum infusion, or vacuum-assisted resin injection)

Instead of over-pressure, we can also employ under-pressure, meaning partial vacuum. This offers the immediate benefit that we only require one rigid mold half: the other half can be replaced by a simple, cheap foil that gets sucked onto the part—in other words, a "vacuum bag." However, thorough resin degassing to remove air bubbles is a must, and mold sealing gets challenging. If we apply a pressure inside the mold of, say, 0.2 bar and have ambient air pressure outside, then even the tiniest air bubbles that seep into the mold will be blown up fivefold, causing inhomogeneities and possibly also surface defects in the part. So an effective seal is needed. A practical and common solution is to surround the part and runner with two seals and to apply a pressure of, say, 0.1 bar between them. With such a set-up, any air bubbles inside the mold will be sucked into the cavity between the two seals (Figure 10.6). The added benefit is that the area between the seals also helps to keep the mold closed, even when the actual mold

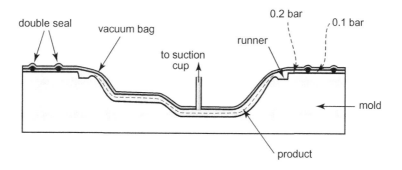

**FIGURE 10.6**

Resin infusion set-up (cross section of mold). (For color version of this figure, the reader is referred to the online version of this chapter.)

cavity is nearly filled and, hence, generates no closing force. The clamping systems that the previous two methods require are not needed.

   Resin infusion can be used for small parts, but its real strength lies in manufacturing much larger items, such as composite boat hulls, wind turbine blades, and even entire bridges. The pressure difference is constant, but because the foil is very flexible, local overlaps of fiber patches do not cause undue variation in permeability as they would in pressure-controlled RTM with rigid mold halves, making the injection time quite predictable. The method's main drawback, apart from the limited pressure difference available (typically just 0.8 bar), is that only one side of the part gets a good surface quality (i.e., the one that is in contact with the rigid mold half). Also, the foil cannot normally be reused, generating waste that is difficult to recycle effectively.

## Light RTM

A relatively young but promising variant to resin infusion is *light RTM*. Basically, this is the same as the previous method, but now the foil is replaced by a thin shell (often made from the composite material itself by hand lay-up over a wooden positive). This shell is sufficiently flexible to maintain the good control over the injection time that vacuum infusion has, yet it is sufficiently stiff to guarantee good surface quality on both sides of the part. In most other respects, this technique is comparable to the previous method.

---

**EXERCISE 10.21**

Which method offers the most design freedom? Think of functional aspects such as shape, size, and production volume, but also of fiber volume fractions.

---

**EXERCISE 10.22**

Which method offers the highest production speeds? And which one has the lowest investments? NB. Even though the methods are generally used for quite different parts (e.g., with respect to size), aim to answer these questions for comparable applications.

---

**EXERCISE 10.23**

Which method offers the best quality? Consider in particular the aspects of roughness, surface defects, and reproducibility.

---

**EXERCISE 10.24**

Can you identify which method was used to make your product examples (Exercise 10.1), and what kind of composite (resin and fiber type) was used?

From the theory discussed so far, we can easily derive design guidelines. For instance, d'Arcy's equation teaches us to minimize the injection length in order to maximize productivity. Practical considerations lead us to minimize the number of suction cups (and preferably to place them on the non-visible side of the product, where the marks they leave behind are less of a problem). However, just as for the other manufacturing principles, such guidelines are only really effective if they are developed for a certain composite (glass-polyester, carbon-epoxy, etc.) and a certain method.

---

**EXERCISE 10.25**

Propose one design guideline, aimed specifically at a certain combination of material and method. Explain your guideline on the basis of the theory covered in this chapter, or common sense. Has it been used to design the product examples you found? If so, how?

---

## Some closing words on costs

As for so many processes, it is not possible to give *the* costs for RTM parts and products, except for specific situations. This is not only because the material costs per kg show a wide range (see Table 10.1), but also because the costs of molds and any other investments for the same combination of base materials, product, and manufacturing method vary widely as well. Given sufficient time and resources, it is possible to automate not just resin injection and curing but also the various preparatory steps, such as fiber patch cutting and mold preparation, as well as the final steps of part removal and edge trimming. Even fiber placement can be automated, which is costly to set up but gives substantial improvements in tolerances, reproducibility, and other quality attributes. This way, even small products (e.g., brake levers for racing bicycles) can be made with very high accuracy and production volumes in the many thousands. If full automation is not feasible, then at least some smart tools and jigs can be made to facilitate manual labor. At the other extreme lies an almost fully manual process, with much lower investments but much higher labor costs (and sometimes poor quality). In practice, this means that RTM costs can really only be determined meaningfully at the level of the manufacturing equipment.

---

## 10.7 Further Reading

In this chapter we focused on RTM, and it is true that there are plenty of other high and low technology ways to process fiber composites. The following sources listed will set you on your way to explore these other processes, as well as to further investigate RTM itself. Furthermore, we have not discussed composites based on natural materials (i.e., those consisting of natural fibers), such as flax or hemp, or bio-based resins. Traditionally, these composites have only been able to compete with materials such as glass-polyester in niche applications, in which their bio-based nature was a pre-requisite to success, on sustainability grounds. However, there are indications that this is about to change, as more bio-based resins are now entering the market. Even more interesting, the natural fiber supply industry, which traditionally was geared toward making fabrics for clothing, is slowly but surely beginning to supply fiber preforms that are optimized for mechanical performance, with aligned instead of woven fibers. Nature-based composites have not yet caught up with conventional composites, but the race definitely is on!

Several simulation software packages exist for RTM, one of which is RTM Worx, available via www.polyworx.com. However, the usefulness of these tools is limited, mainly because permeability values are difficult to find. While designing and developing RTM products, we should also rely on common sense, experience, and testing—which is, of course, how we should approach design for manufacture in general: no amount of theoretical insight can fully replace what just a few days in a workshop can teach you.

There are many journals in this field, with differing degrees of emphasis on design and processing, a good example being the *Journal of Reinforced Plastics and Composites*.

The world's largest industry fair in the field of composites is the JEC, held annually in April in Paris, France (see www.jeccomposites.com for a wealth of information on all aspects of the composites industry, including design and manufacture).

There are many well-written books that deal with composites. We can recommend these three:

Beukers, A., van Hinte, E., 1998. Lightness. 010 Publishers, Rotterdam, NL. (*A must-read for every designer. Covers structural design, materials, and manufacturing.*)

Braddock, S.E., Mahony, M.O., 1999 and 2008. Techno Textiles, vols. 1 and 2—Revolutionary Fabrics for Fashion and Design. Thames & Hudson, London, UK. (*Inspiration from quite a different angle—the world of fashion and high-tech fabrics.*)

Miracle, D.B., Donaldson, S.L. (Eds.), 2001. ASM Handbook, vol. 21: Composites. ASM International, Materials Park, Ohio, USA. (*One of the key reference books, covering virtually all aspects of the field from design to manufacture, maintenance, and recycling. Expensive, but—as with all ASM Handbooks—worth the investment.*)

# Additive Manufacturing

This saxophone mouthpiece was made using additive manufacturing, allowing a geometry that would be difficult to make with other processes at competitive cost and time-to-market. It was used to great effect at the 2012 North Sea Jazz festival.

**Manufacturing and Design**

**187**

## 11.1 INTRODUCTION

*Three-dimensional printing makes it as cheap to create single items as it is to produce thousands and thus undermines economies of scale. It may have as profound an impact on the world as the coming of the factory did.... Just as nobody could have predicted the impact of the steam engine in 1750—or the printing press in 1450, or the transistor in 1950—it is impossible to foresee the long-term impact of 3D printing. But the technology is coming, and it is likely to disrupt every field it touches.*

*The Economist,* **February 10, 2011**

Additive manufacturing, including the processes known to the general public as "3D (three-dimensional) printing," is the youngest of all manufacturing principles covered in this book, and potentially the most disruptive. Key to the process is that it builds up parts gradually by the addition of material (typically, but not exclusively, in layers) under full digital control. The input for a 3D printer is the raw material plus a computer-aided design (CAD) file, the output is a part ready for finishing. The range of available materials is increasing steadily, from low-quality brittle plastics some years ago to high-quality engineering polymers, metals, and ceramics today. Form freedom is virtually unlimited. Moreover, each part can be unique, as part-specific molds or dies are not required. All that is needed is a different CAD file. Furthermore, with the equipment for additive manufacturing dropping dramatically in price, the process becomes more and more available for the increasingly computer-literate public. "Fablabs" are opening up around the world to accommodate a growing range of professionals, flanked by online 3D print shops where consumers can order parts of their own design.

3D printing currently receives considerable attention in the media, but perhaps surprisingly, the breakthroughs and patents underlying today's developments date back to the 1980s. So the revolution—if that is the proper term—has been going on quietly for longer than most people might assume. Will it eventually be as disruptive to the field of manufacturing as *The Economist* anticipated in 2011? That remains to be seen. What is certain is that the process has already firmly secured its place in industry, and no designer can afford not to know about it. This chapter acquaints you with the basics and the unique capabilities that additive manufacturing offers, along with its key challenges and limitations, which do not regularly get the attention they deserve.

---

**EXERCISE 11.1**

Run several Internet searches with different time ranges (e.g., for 2010, 2011, and 2012), using both "3D printing" and "additive manufacturing" as your keywords. Each time, compare the top five search results. What, or who, are these?

---

## 11.2 ADDITIVE MANUFACTURING AT-A-GLANCE

Figure 11.1 shows the general principle of additive manufacturing. A part gets built up layer by layer, with layer thicknesses typically being some 100 μm, until it is ready. Exactly how this can be done, and with which materials, is discussed in the next section; for now, we can immediately

**FIGURE 11.1**

General setup of additive manufacturing (supporting material: white).

make the important observation that even if the process may be quick to set up, it is slow to run. We can also observe the manufacturing triangle and its trade-offs by noting that speed goes down with resolution: the smoother the part has to be, the thinner the layers will get, and the longer that manufacture will take. Progress in this field has been amazing, with printing speeds almost doubling every few years and resolution getting steadily better, but when it comes to per part speed, the process will never catch up with, for example, injection molding or thermoforming. On the positive side, virtually *any* shape can be made, as each layer can be different from the next one. Form freedom can be greatly expanded by using support structures, which allow higher layers to extend away from lower layers—in other words: hollow parts and parts with undercuts (again, see Figure 11.1). We can even make multi-body mechanisms in a single operation, such as working sets of interlocking cogwheels, "chain mail" fabrics consisting of interlaced rings, lattice structures, and so on. Depending on the exact process, it may be possible to mix different materials together into one part, including depositing a continuous gradient in composition. So the material itself becomes a design variable.

---

**EXERCISE 11.2**

With typical layer thicknesses of 100 μm (although it can be as thin as 16 μm in extreme cases) and a printing speed of 10 sec/layer, roughly how long would it take to produce an ordinary coffee cup?

---

After finishing, the support structures must be removed, and depending on the method and equipment, this can be time-consuming. In general, we must anticipate that parts coming straight out of the printer have a coarse grainy surface texture and will require some finishing if they are to be smooth and fully functional. For instance, a 3D-printed snap-fit joint will not work as well as one made by conventional injection molding, unless its working surfaces are carefully finished.

The workflow for additive manufacturing is as follows. First, a CAD file is made of the part in question and exported as a temporary file in the "standard tessellation language" (STL) format. This means the part surface (i.e., its boundary) is represented by triangles, as shown in Figure 11.2. Smaller triangles provide a more accurate approximation of the original CAD file, but at the expense of file size, which may in extreme cases slow down operations but is normally not a problem. The STL file is then sent to a suitable 3D printer and the part is manufactured. Support structures, if these need to be defined separately, are usually generated by the 3D printer's embedded software, facilitating the process. The original CAD files are easily stored and can be parameterized so that the designer can easily vary the length, height, or other characteristics of the final part, all based on a single file. This is partly how additive manufacturing can be exploited to "customize" parts.

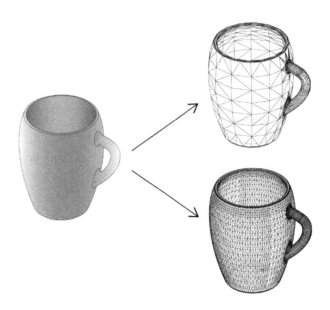

**FIGURE 11.2**

CAD model meshes of a coffee mug (top: 3,000 faces, bottom: 13,000 faces).

---

**EXERCISE 11.3**

For additive manufacturing, STL has another, somewhat older meaning. What is this?

---

## 11.3 SUB-PRINCIPLES AND MATERIALS

Standardization of the processes and terminology for additive manufacturing only took place recently. In fact, even the term "additive manufacturing" and its abbreviation "AM" only date back to the early 2000s. People still refer to variants of the same process by different names: for example, what the Stratasys company calls "fused deposition modeling" (FDM) is basically the same as what 3D Systems calls "plastic jet printing," whereas others refer to it as "fused filament fabrication." Needless to say, this only breeds confusion, with some trademarked terms referring to a specific company or printer, whereas other terms have more general applicability. Here we will adopt the standardization and terminology laid down by the American Society for Testing and Materials (ASTM) in 2012. Seven variants are distinguished here, which under the system of this book are best considered as AM sub-principles, as most of them can be done with quite different methods and equipment:

1. Powder bed fusion
2. Vat photo-polymerization

3. Material jetting
4. Material extrusion
5. Sheet lamination
6. Binder jetting
7. Direct energy deposition

The first and perhaps most important sub-principle, at least from the perspective of industrial manufacture, is *powder bed fusion*. Currently the main industrial player behind this process is EOS, which refers to its process as "direct metal laser sintering." Another name for the same process is "selective laser sintering" (SLS), which is still commonly encountered. Figure 11.3 shows how it works. A powder bed fusion machine operates by depositing a thin layer of powder onto its worktable and using a laser beam to heat this powder where it must be fixed, sintering the particles together (although melting is also possible; see Section 11.6). This is done in an enclosed operating volume. Once the first layer is finished, a second layer is deposited over it, which is also selectively sintered, and so on, until the part is ready. During the process, unsintered powder remains in place, supporting higher layers. This material is easy to remove afterwards and can be reused, although not indefinitely: the partially sintered particles close to the sintered contours must be separated to avoid loss of processability and part properties, and are therefore lost. Powder bed fusion can use many different materials, including thermoplastics, metals, and even ceramics, although this obviously requires very different pieces of equipment. Currently, it cannot yet combine materials (e.g., different types of plastics) in one component, yet this may change in the near future. Parts tend to have a grainy, rough, and somewhat porous surface; metal parts must usually be post-cured or impregnated to reduce porosity to acceptable levels. Although the mechanical properties of the sintered materials allow for load-bearing applications, they do not yet match those of, for example, molded plastics or forged metals, especially in terms of ductility and toughness.

The second sub-principle is *vat photo-polymerization*, also commonly known as stereo-lithography (SL, or STL). This was actually the oldest AM process, with commercialization dating back to 1986. It is comparable to powder bed fusion, but instead of using a laser to heat and sinter solid powder particles together, vat photo-polymerization uses an ultraviolet (UV) laser to cure a UV-curable monomer dissolved in a solvent. So it is a liquid-deposition process, using thermosetting polymers and rubber-like materials. Process control and extraction of the uncured material require it to take place in a controlled environment (i.e., inside a "vat"). Support structures can be printed along with the part itself to support overhanging layers; these supports, often shaped as honeycombs, must be mechanically removed

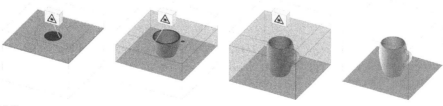

**FIGURE 11.3**

Schematic of powder bed fusion.

afterwards (i.e., by breaking them away and sanding the fracture points) and cannot be reused. Compared to powder bed fusion, vat photo-polymerization gives a better surface quality because the polymer's fluid behavior tends to blur the individual layers; it also allows for somewhat larger parts and for equal size and speed, and the equipment tends to be cheaper as well.

---

**EXERCISE 11.4**

What are your expectations of the material properties (strength, stiffness, ductility) of vat photo-polymerized parts made of thermosetting plastics? What about rubbers?

---

In *material jetting*, individual droplets of material are deposited layer by layer (Figure 11.4). Currently, the main company behind this process is Objet. Its PolyJet machines typically have two print heads: one for the actual part and one for support structures. The main material is a viscous thermosetting resin that is cured by a high-intensity UV light source as it is being deposited, whereas the support can be a water-soluble substance, allowing easy removal. Like vat photo-polymerization, material jetting allows for relatively smooth part surfaces. A key benefit it offers over other commercially available AM processes today is that it can also combine different materials into one part (e.g., thermosets and rubbers) with discreet or even gradual transitions from the one to the other and back. The machine then has three print heads: two for the materials, one for the supports. Material properties are sufficient for load-bearing applications, but ductility (for thermosets, not rubbers) is again low.

**FIGURE 11.4**

Schematic of material jetting.

---

**EXERCISE 11.5**

Search the Internet to find the largest part sizes that the three sub-principles mentioned so far can print. Make sure to compare only equipment for polymers. Which one "wins"?

---

The fourth sub-principle, *material extrusion*, uses a wire as its input, which is extruded through a heated nozzle and fused together by cooling in ambient air, gradually shaping the part (Figure 11.5). First, the machine deposits one layer of material by moving the nozzle in the horizontal plane to trace the required layer shape (turning the supply of material on and off as needed), then moves the nozzle up along the vertical axis, repeating the procedure until the part is finished. This *material extrusion* process is known by several other names, including fused deposition modeling and plastic jet

**FIGURE 11.5**

Schematic of material extrusion.

printing. The input for commercially available equipment consists of thermoplastics (typically ABS, PC, and PLA), supplied on a spool or reel. To some extent, the process allows higher layers to overhang lower ones; for deeper undercuts and additional form freedom, "break-away" support structures can be printed much like those used in vat photo-polymerization (with clever process control to prevent bonding between part and support, facilitating support removal). Some equipment can also fabricate water-soluble support structures using a second nozzle.

Compared to the previous sub-principles, material extrusion is well-suited for desktop work: it does not involve fine powders, intense light or heat, or reactive chemicals; indeed, the process can take place in ambient air (although accurate parts require an enclosed, heated chamber to minimize part distortion). This ease of use, plus the fact that the cost of 3D printers using material extrusion has dropped dramatically, are strong drivers for the current do-it-yourself revolution. For the manufacture of substantial batch sizes, material extrusion is less well-suited than the previous sub-principles, mainly due to its lower speed (at equal resolution). Material properties, as usual, are not quite comparable to those of conventionally processed plastics.

Most sub-principles covered in this section are developing rapidly, but progress in material extrusion is perhaps the most stimulating for the imagination. For instance, processing of metal wire is currently being explored and may soon become commercially available. Conventional inert gas welding technologies with multiple wire feeds provide another experimental technique for AM, a process that allows deposition in any orientation, including building material out horizontally, or beneath an overhang. The scope for both is greatly expanded by mounting the extrusion nozzle (or welding torch) onto a conventional robot arm. Further unusual developments include replacing the wire spool in extrusion with a reservoir of foam or even concrete. Methods for 3D printing can be quite varied indeed!

**EXERCISE 11.6**

What are the base prices and melting temperatures for aluminum, magnesium, tin, and zinc? Which one(s) would be suitable for home applications of AM using material extrusion, based on these data?

*Sheet lamination*, also known as laminated object manufacturing (LOM), is the fifth AM sub-principle. Its input is a roll or stack of thin sheet material, coated with an adhesive. Successive layers are cut out in the required shape and stacked to build up the part. Input materials can theoretically be paper, plastic, and even metal. Compared to the processes covered so far, sheet lamination is less restricted in part size. However, surface quality is poor, and the process generates a lot of waste (so it is "subtractive" as well as additive). The fact that the only commercial equipment available today is limited to paper does not help to increase application beyond non-structural prototypes and mockups.

---

**EXERCISE 11.7**

As a percentage of the input material, how much waste would you have if you made a coffee cup using sheet lamination?

---

In *binder jetting*, a liquid adhesive is selectively deposited to join powder materials, typically gypsum or starch, which are themselves deposited in thin layers, much as in powder bed fusion. The key commercial player pushing this process today is ZPrinter. Parts made with its equipment are only semi-structural but can have any combination of colors, ideal for mockups of products or scale models of buildings. The term "3D printing" was, in fact, originally used for this process only. Under the same sub-principle, ProMetal is a similar process for metals. Metal powder is joined by binder jetting to produce a "green part," which is sintered in a furnace and then infused with a secondary metal.

Finally, in *direct energy deposition*, both the material (usually a metal powder) and the energy needed for fusion (usually laser heating) are simultaneously focused at the same location. The metal powder is projected onto a surface, where the focused laser directly melts the powder in place. Material properties are very good, without porosity. Because this process does not involve a bed of material, it does not require printing in horizontal layers. Therefore, the equipment can also be placed on a robotic arm, which allows printing onto existing objects (e.g., to coat or repair a part). The leading industrial player behind this process is Optomec, which is also working on 3D printing of electronics under the name of Aerosol Jet.

---

**EXERCISE 11.8**

Make an AM overview. Which of the seven sub-principles are (currently) suited for metals, which for plastics, and which for both? Which allow small-series production, and which are more suited for prototyping? What kind of part finishing is needed?

---

It is important to note that the form freedom offered by additive manufacturing goes well beyond the ability to produce parts with undercuts, and even further than making mechanisms. The material microstructure and composition itself become design variables, allowing, for example, the manufacture of regular foam-like or lattice materials, structures with integrated light guides, graded materials, and so on—there is a whole world of possibilities to explore. Additive manufacturing is still developing rapidly, with the limits of the sub-principles, in terms of function, quality, and cost, constantly being pushed back. We can, however, safely predict that for industrial manufacture of significant batch sizes, *powder bed fusion* ("SLS") and *vat photo-polymerization* ("STL") will continue to be the main processes for several years to come. Moreover, AM has several challenges and problems that are not likely to disappear soon, and it is to these that we now turn our attention.

## 11.4 DIGGING DEEPER: CHALLENGES AND PROBLEMS

A clear example of interaction within the manufacturing triangle has been mentioned: the relatively low speed of additive manufacturing, and the trade-off between speed and roughness inherent to the principle. Depositing thin layers gives a smoother finish but is slow; conversely, faster deposition gives a rougher finish. Here, we discuss four more challenges and problems: material anisotropy, difficulties associated with varying the input CAD file, issues associated with part certification, and material cost. Although the details will vary from one method to the next (and among different equipment sets), these problems are largely inherent to the principle, and every designer should be aware of the issues.

First we consider material anisotropy. If we analyze, say, a part made by powder bed fusion, we find that its mechanical properties (strength, toughness, ductility, etc.) within the horizontal plane differ from those in the vertical direction. This suggests that the sintering contacts between powder particles evolve differently in these two directions. For the other sub-principles, the mechanisms behind the anisotropy are different, but the result is the same: mechanical properties *within* layers vary from those *between* layers. Figure 11.6 shows what this can mean. Obviously, the industry is hard at work to minimize these differences, but they are inherent to additive manufacturing and can therefore never fully go away.

**EXERCISE 11.9**

From Figure 11.6, estimate the (average) tensile strength and ductility in both planes. How large are the differences? And how do these properties compare against those of a standard injection-molding grade of PA?

The issue of anisotropy gets extra weight once you realize that multiple copies of a single part need not always be printed in the direction that its designer intended. Commercial suppliers of parts often print several different parts simultaneously to save time and cost. After all, for most AM processes, the

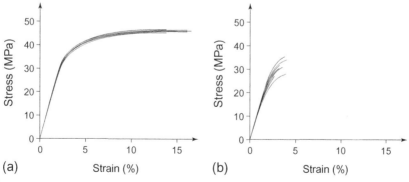

(a)     Strain (%)          (b)     Strain (%)

**FIGURE 11.6**

Stress-strain curves for powder bed fusion deposited nylon (PA), for multiple test samples oriented in (a) the horizontal plane and (b) the vertical plane. (For color version of this figure, the reader is referred to the online version of this chapter.)

*(Courtesy: BPO, Delft, The Netherlands.)*

material deposition steps last equally long regardless of how many different parts you put together in the machine; it also reduces the amount of powder that must be reused. Smart software is deployed to "nest" the different parts in three dimensions and make optimal use of the machine's operating volume, and depending on the other parts in the batch, one copy of a certain part may be oriented along one axis in one batch and along another axis in the next. Consequently, the material properties will vary as well. Controlling this variation would require the AM workflow to take the material properties into account, but this is currently not the case—the STL file format only defines the part surface.

The next challenge lies in making and varying the input CAD file. A key strength of AM is its ability to make each part unique. Indeed, if the principle is to make a significant impact and compete with established processes for large production volumes, this strength *must* be exploited, because additive processes cannot compete on time and cost alone. For instance, a 3D-printed bicycle helmet will not be able to compete against one made by thermoforming, unless it can be custom fit, offsetting the higher cost of manufacture with improved functionality and, hence, value. However, this implies that it must be possible to make unique versions of the CAD file easily and cheaply. If this preparatory step takes too much time, then the end result will still not be competitive. Returning to the example of the bicycle helmet, current 3D scanning technology can already be used to scan the heads of customers quickly, but transforming the scan data into a suitable CAD file still takes considerable effort. All kinds of custom-fit products can be made this way, but the challenges involved in making and varying the input file should not be under-estimated. Solutions are continuously being sought, but the problem is far from being solved.

Incidentally, this issue points to a new role for designers: instead of making a single, fixed design, they are now required to come up with configurable, parameterized designs. In other words, a *process* has to be designed-in simultaneously.

---

**EXERCISE 11.10**

Estimate the extra cost of making the CAD file, assuming it takes 15 minutes for a skilled operator to make one. How does this affect the product retail price?

---

The third challenge concerns part certification. To explain this issue, we first recall the difference between safety-critical and non-safety-critical products. The former are parts and products whose failure can put people at risk of injury or death, for example, the brakes of your bicycle. The latter are objects where failure is annoying but not hazardous, for example, your bicycle saddle cover. Safety-critical parts must be designed such that catastrophic failure is either extremely unlikely (the "safe life approach") or, if this is not feasible, such that failure is gradual instead of catastrophic and backed up by other components (the "fail-safe approach," with multiple load paths and redundancy). In both cases, the variability introduced by the materials and manufacturing processes must be kept under control, so that even in a worst-case scenario, deviations from the target properties and tolerances cannot unduly reduce part integrity. Suppliers can, of course, be held accountable for quality problems. So in industries where safety is paramount, such as in the manufacture of cars, aircraft, or pacemakers, it is therefore standard practice to only use materials and manufacturing processes that deliver certified performance.

Certification involves extensive part testing under controlled circumstances, according to industry-defined standards, then manufacturing the real parts under approved process settings, conducting statistical quality assurance checks on batches of the output parts. Input materials are also regularly tested against the relevant standards. This approach to certification works well when you make large numbers of identical parts, and it is deployed in essentially all industries. For additive manufacturing, however, it is poorly suited, as a key strength of this process is the ability to make each part unique. Testing every part is much too expensive in all but the most extreme situations. Alternative pathways toward certified performance are being explored (e.g., by organizations such as the ASTM and International Organization for Standardization [ISO]), but this is another AM problem to be solved. Consequently, it is currently still difficult to use AM parts for safety-critical applications.

---

**EXERCISE 11.11**

What would be the market value of a custom-fit 3D-printed bicycle helmet that has no certified performance?

---

Finally, we get to cost. Although 3D printing may involve no part-specific investments (at least, if we exclude the cost of making the input file), with machines costing thousands, even hundreds of thousands of euro, the machine cost can be substantial. Even more important, the materials are expensive. Polymer input for powder bed fusion and vat photo-polymerization typically cost 30 to 50 euro/kg today (more for material jetting, somewhat less for material extrusion). There is no fundamental reason why these prices are so high—indeed, they should eventually get close to the 3 to 10 euro/kg for ordinary engineering plastics, such as ABS, PA, and PC—but prices are falling only slowly: a factor of about 2 since the early 2000s. In all likelihood, small batch sizes and narrow tolerances are the main reasons for these high prices (not surprisingly, they also contribute to the current business model of most machine manufacturers, who are often the sole suppliers of the input material). For powder-based input, there are the added problems (and associated costs) inherent to handling fine powders—recall that many plastics, including PA and PC, are somewhat hygroscopic.

For metals, the prices are far higher: for powder bed fusion, the metal input materials are currently at least 10 times more expensive than polymers, easily topping 500 euro/kg. The high cost is due to many factors: the process commonly uses metals that are expensive to begin with (such as titanium), it is used in relatively small batches with stringent quality demands, and excess powder cannot be reused indefinitely. Furthermore, producing and working with ultra-fine powders is inherently more costly, because metal powder handling, from production, to storage and transportation, to application in the 3D printer, must be done under an inert atmosphere (e.g. under Argon gas), otherwise excessive oxidation would lead to poor material properties. Depending on the metal in question, ultra-fine metal powder can also represent a fire hazard.

---

## 11.5 APPLICATIONS

Since the 1980s, additive manufacturing has moved first hesitantly, then steadily ahead from the laboratory into industrial practice. New applications are being announced almost daily, and it is impossible to say how the principle will eventually unfold. And each sub-principle has variants in methods, such as material extrusion with and without enclosed volume for temperature control or with and without an extra nozzle for support fabrication. As a result, it is somewhat pointless to discuss the process limitations of these variants (relating to size or speed or materials), as the information will quickly become outdated. Instead, we will focus on the main application areas to watch out for.

## Rapid prototyping

The ability to quickly and comparatively cheaply make accurate models is highly useful in any design setting. Now that 3D printing materials can obtain properties that are comparable to those of the materials used in actual parts, 3D printing is becoming indispensable as a means for prototyping. Instead of waiting for weeks for handmade models, you can have them in your hands overnight—at much lower cost and generally with superior fidelity compared to what even the best model makers can deliver. What's more, 3D printing allows you to go through more iterations of your design in the same development time, getting an optimal result much sooner. For the same reason, it enables more complex designs, as there are more chances to get things right. In rapid prototyping, designers often start with material extrusion or sheet lamination for the first attempts, then they move up to more expensive but more capable processes, such as powder bed fusion or vat photo-polymerization, for higher quality prototypes.

It is also interesting to note that under the heading of "rapid prototyping," there are several non-additive processes as well, notably integral machining, laser cutting, and 3D bending (e.g., of metal tubing). Like AM, these processes allow the manufacture of a unique part using only a CAD file as specific input, and in principle they enable the same rapid delivery expected for prototyping. Increasingly, supply chains are being reconfigured to meet the new expectations, mimicking that of a 3D print shop. This indicates the disruptive impact of 3D printing, alluded to in the quote from *The Economist* at the start of the chapter, forcing established technologies to up their performance.

## Rapid manufacturing and mass customization

The incredible form freedom of additive manufacturing allows you to make highly optimized parts that are prohibitively expensive, if not impossible, to make with other processes. In such instances we speak of "rapid manufacturing"—something of a misnomer, as the process itself is anything but fast. But it does not need to be, given the key benefits: the design freedom and with it the potential for shape optimization, the short time to market, the ease of making improvements, plus the absence of capital investments. A good example is set by plastic air ducts for helicopters. These parts need to be very light, have intricate forms, and are made in limited numbers (e.g., 10 to 50 units/year). Or think of medical applications, such as metal implants. Collimators of X-ray equipment, which are the parts that narrow the X-ray beams, are now also made using AM, depositing tungsten in a geometry that would be impossible to make with other processes and significantly improving performance. But outside the medical and aerospace sectors (that is, outside of high-value environments), applications are still rare, and many rapid manufacturing applications announced on the Internet often turn out to be still at the experimental stage. This also goes for the much-talked-about "mass customization" of products, which is indeed promising but is held back by the challenges discussed in the previous section.

One exception to this rule is the use of additive manufacturing to temporarily supply parts when the definitive parts are not yet available. This occasionally happens in the automotive industry, where the development times are short and the stakes are high. For instance, if a mold for injection molding is not yet available because its development was rushed, it is now possible to substitute 3D printed parts for the first production runs—though the parts in question must be tested to check that they meet the specifications. Also, as maintenance of new cars is routinely tracked under warranty, it is possible to replace the temporary part with its final version later. Even so, this approach is not yet used for safety-critical components.

A second exception, and one that is of increasing importance, is the use of additive manufacturing to supply spares. This has a huge potential to save costs by eliminating the need to manufacture and store the hundreds, sometimes thousands of spare parts necessary to maintain complex products. Also, 3D-printed spares can be made on the spot, speeding up repairs. The first successful research projects into this promising application have already been executed and although the issue of certification complicates things, we can surely anticipate more of this in future. The U.S. military is now using this approach to facilitate maintenance, and others will surely follow. Indeed, in June 2013 there was news that the National Aeronautics and Space Administration (NASA) plans to bring a 3D printer on board the International Space Station.

## Manufacture of molds and dies

Another area of application that no designer should overlook is the use of additive manufacturing to make molds and dies (i.e., to make the part-specific investments for conventional processes such as plastic injection molding or metal casting). Here, the form freedom of 3D printing can be deployed to add complexity that is not available to conventional tool-making processes, such as the inclusion of complex, curved cooling channels. This too is still in the experimental stage, and conventional mold materials and manufacture are unlikely to be fully replaced by the combination of metals and, for example, powder bed fusion, but it is worth watching the developments in this field.

A more accessible use of 3D printing in relation to mold making, which is becoming quite common, is to use it for making casting patterns. For example, material extrusion can be used to make the wax positive for investment casting of metals, or vat photo-polymerization can be used to make the plastic pattern used for the molds in sand casting of metals.

## DIY home manufacture

The final application to follow is the do-it-yourself world of home manufacture. With prices for entry-level desktop 3D printers (usually based on material extrusion) dropping below 1,000 euro, sales of these machines are soaring. In parallel, an array of support products and services is being set up, such as libraries of CAD files of popular parts and products, easy-to-use CAD programs to make your own designs, and even simple 3D scanners to copy the geometry of existing products. The result is a rapidly growing community of people using 3D printers at home. This has already led some enthusiastic commentators to speak of "democratizing manufacture," whereas others have voiced concerns about how all that 3D-printed material will be recycled. Of course, the means of manufacture have always been available to any sufficiently determined individual (e.g., a sowing machine or a small lathe), but the high accuracy of 3D printing and its digital nature, which allows people to link into an online community and removes the pre-requisite to possess significant manufacturing skills, mean that "making things" is a lot easier than before. But the actual *designing* of those things remains something else, requiring skills, knowledge, and creativity that the general public does not possess, so the impact will likely be less than many trend watchers expect.

At any rate, it is not in the cards that conventional manufacture will be taken over any time soon. The same challenges that hold back many other applications will also limit DIY home manufacture. More likely, this application will come to complement the established manufacturing opportunities, much as AM is doing in the other applications discussed earlier.

## 11.6 Further Reading

As additive manufacturing remains an emerging principle, this introductory chapter on the topic has taken a different approach to the earlier core chapters, presenting a critical view on the main innovations, problems, and challenges. We have been necessarily brief in our coverage of the seven sub-principles. For instance, powder beam fusion extends to methods that melt the powder instead of merely sintering it, giving better material properties (e.g., "selective laser melting" and "electron beam melting"). And we have not looked into 3D printing of foodstuffs, which some believe may revolutionize cooking and eating, or processing of living materials (such as cells), which is already at the experimental stage in medicine. Likewise, our attention for part optimization has been minimal: here, something to watch are "topological optimizations"—that is, programs that automatically generate ideal shapes inside a predefined volume under a given load case, mimicking the way that natural growth processes operate in bone or wood. With suitable post-processing, such shapes can then be 3D printed, often yielding substantial weight savings. Then there is the combination of 3D printers with robotics, which holds great promise for flexible automation of manufacture and the assembly of complex products. Direct manipulation and grading of material microstructures and compositions is another subject to keep an eye on, allowing greater material customization than before.

Academic publications are still scattered, which is logical for a field that is not yet consolidated. Currently, the main journals are *Rapid Prototyping Journal, Journal of Materials Processing Technology, International Journal of Advanced Manufacturing Technology,* and *Virtual and Physical Prototyping.* EuroMold is the worldwide leading trade fair for mold making and tooling, held annually in Frankfurt, Germany (visit www.euromold.com). This large event also gives ample attention to additive manufacturing, from processes and equipment to materials and applications.

Perhaps more than is the case for most other processes, AM lends itself to learning by doing: make a CAD drawing, get it 3D printed, and see what you have made! Staying abreast of developments is best done via the Internet (and, of course, by visiting companies), as books on the topic tend to be outdated the moment they get in print (hence, the cautious approach to defining process limits that we have taken). But one recommended source is the following:

Gebhardt, A., 2012. Understanding Additive Manufacturing. Hanser, Munich, Germany. (*A book that is no-nonsense, in-depth, and informative.*)

# Joining and Assembly

# 12

Recent developments in joining include this precipitation-hardened martensitic steel bolt. Combining ultra-high strength with affordability, it has been applied in truck engine cylinder heads since 2013.

*Image courtesy Koninklijke Nedschroef Holding.*

## 12.1 INTRODUCTION

Joining is the process of putting two or more parts together, whereas assembly is the term used to describe the sequence of multiple joining operations involved in making a product. For many good reasons, nearly all products consist of different parts, involving a variety of joining processes. A standard bicycle, for instance, has a brazed steel frame onto which all other components are screwed, welded, and clamped. In turn, most of these components—saddle, handlebars, brakes, and so on—also consists of multiple parts that have been joined together. Assembly not only allows us to combine different materials and parts into products, but also to make different *versions* of the same product: think of the range of frame sizes we can combine with standard sets of the other components, or think of the range of bicycles with different specifications, but all built on the same frame.

Joining and assembly can account for up to 30% of a product's total manufacturing cost and up to 50% of its total manufacturing time. If that does not sufficiently underline the importance, then consider that joining can impact considerably on product performance, for example, by introducing stress concentrations or by modifying the microstructure and properties (sometimes catastrophically: many failure analyses trace their origins back to joints). Assembly may come toward the end of product manufacture, but it is potentially an expensive mistake to leave it to the end of the design process—if only at the basic level of ensuring access to the joining equipment on a production line. Joining and assembly problems have been the Achilles' heel of many designs and products.

For high-volume products, such as cars and consumer electronics, the processes and operations involved have been refined to a high level. This does not necessarily involve high investments, and many of the smart assembly solutions developed for those products are equally applicable to more modest production volumes. But joining is not just about functionality, cost, and speed: joints often contribute to the character and "look and feel" of products by expressing a certain style and invoking a perception of quality. For instance, neat rows of smooth rivets tell a different story than that told by rough weld beads or near-invisible adhesive joints (see the examples in Figure 12.1). There is a clear design potential here waiting to be explored that goes beyond purely meeting technical requirements.

Just as there are many different shaping principles and methods, there are numerous joining processes—in fact, joining is best divided into five sub-principles:

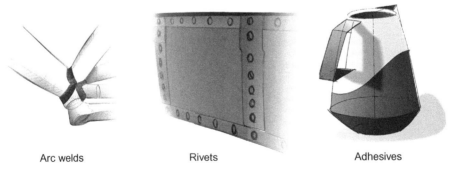

Arc welds            Rivets            Adhesives

**FIGURE 12.1**

Examples of joining look and feel.

- Welding (suitable for most metals and plastics)
- Brazing and soldering (certain metals)
- Adhesive bonding (nearly all materials)
- Mechanical fastening (nearly all materials)
- Joining using form closures (nearly all materials)

Apart from representing distinct industrial branches, each sub-principle has a common physical basis in how the process works, and it can then be broken down into various methods and the different types of equipment used. To keep this chapter concise, we limit ourselves to the methods you are most likely to encounter and refer you to other sources for details about other processes (but as usual, you will be equipped to ask the right questions in studying them).

This chapter first presents the manufacturing triangle for joining, along with its various attributes of function, quality, and cost. This provides the necessary terminology, context, and background for a discussion of the five sub-principles, their main joining methods, and finally some comments on overall assembly issues.

In closing this introduction, we note that the boundary between shaping and joining can be fuzzy. For instance, 2K injection molding is a shaping process that integrates a joining step, as does the use of inserts in casting or molding. Co-extrusion of PVC-coated copper wire can either be seen as shaping a simple component or as making a joint between PVC and copper. No simple definition can capture every example, but we can intuitively see when it is important to consider a joining step in its own right.

---

**EXERCISE 12.1**

Look around your house for examples of joining, finding at least one product per sub-principle.

---

## 12.2 THE MANUFACTURING TRIANGLE FOR JOINING

Of the five functional attributes listed in the manufacturing triangle (see Figure 1.3 in Chapter 1), it is straightforward to consider size, production volume, and "look and feel" in the context of joining. Material, however, needs reinterpretation: the materials to be joined can be either similar (e.g., carbon steel to carbon steel) or dissimilar (e.g.. carbon steel to wrought aluminum), and the process may also introduce new material (e.g., solder or an adhesive). Likewise, shape acquires additional meaning, not only in terms of the shape of the parts but also regarding how the joint brings those parts together: overlapping, at right angles, and so on. Figure 12.2 presents four common joint geometries, to illustrate the different possibilities. In addition, joints may be required to be water- or air-tight, which also relates to shape, as is the possible requirement (related to look and feel) to be seamless.

One further functional attribute, specific to joining, is closely linked to the environmental impact of the overall product and also to maintenance and life cycle costs. This is whether the joint is permanent or not, and how quickly and easily joining can be reversed. With increasing demands for material recovery at the end of a product life, "design for disassembly" gains importance. Here the trade-off is between the speed (and cost) of separating the parts and the level of contamination of the materials

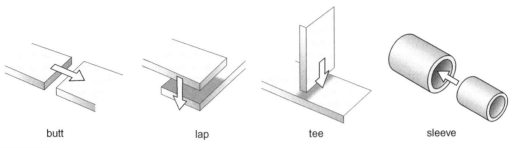

butt                    lap                    tee                    sleeve

**FIGURE 12.2**

Common joint geometries: butt joint, lap joint, tee joint, and sleeve joint.

for recycling purposes. Reversible joining processes also favor modular design, extending product life-times by offering the ability to replace parts of a product without having to scrap the whole.

> **EXERCISE 12.2**
>
> Which of these four joint geometries are involved in the sample products you found for Exercise 12.1? Are the materials similar or dissimilar? And are the joints permanent, or can they be undone (either easily or with some difficulty)?

All of the five quality attributes (tolerances, roughness, defects, properties, and reproducibility) apply to joining. Tolerances relate to the joint itself and to the overall product, especially with respect to how well the parts are aligned and spaced after joining. Such errors accumulate in assemblies of many parts, such as an automotive body, meaning that later joints fail to fit up correctly—optimal sequencing of an assembly process is a design challenge in itself! Joining can modify roughness but is more importantly a source of defects (e.g., porosity, micro-cracks, and, above all, residual stresses). Any thermal joining process is susceptible to the differential expansion and contraction associated with residual stress and distortion. Furthermore, thermal processes may locally melt the material to make the joint or may modify its microstructure and properties in the solid state. The combination of defects and property changes around joints can be of great significance for strength, toughness, fatigue, and corrosion. The history of technology is littered with examples of joining-related product failures, some leading to considerable loss of life and major litigation. In spite of our extensive understanding of materials behavior, joining remains a source of technical difficulty and should not be underestimated, particularly in safety-critical design. Closely related is therefore reproducibility—processes in which the outcome can be sensitive to process and material variables must be controlled within the required specification, each and every time.

Finally, we move on to economic attributes (i.e., part price, investment, time-to-market, social factors, and ecological costs). Because joining is often done within a production line, it can be difficult to break down the cost contribution of specific joining operations. But similar arguments will apply to trade-offs, such as higher investments enabling higher throughput. The sustainability-related issues will mostly be viewed in the context of the product as a whole, rather than for the joining processes themselves. An exception relates to disassembly and recycling, particularly for joints between dissimilar materials. Material contamination is a key aspect of recycling processes—even in joints between identical materials, other materials may be introduced, such as bolts, washers, adhesives, gaskets, and the like. As Chapter 14 will show, full product disassembly is almost never economic during recycling, and the labor costs largely determine how much is removed before the product goes into a shredder. So it is not sufficient just to select joint types that can be undone if they are left intact in the disassembly stage—what then matters is how well the joints "survive" the shredding process.

## 12.3 WELDING

Put simply, welding involves aligning two parts close together and locally heating the interface region to form the joint. In *fusion welding*, the peak temperature exceeds the material's melting point, forming the joint as the weld zone solidifies. *Solid-state welding* avoids melting and uses hot deformation and diffusion instead. Welding is suitable for most metals and thermoplastics. With few exceptions, welding can only join similar materials (e.g., low carbon steel to low carbon steel, or PP to PP). However, it is very flexible in terms of joint geometry and can be used to create either local joints ("spot welds") or continuous ones ("seam welds"). Welds are permanent, can be very strong and, if desired, can be airtight. Some welding methods use just the material from the parts themselves to make the joint, whereas others rely on the addition of a suitable weld filler material to fill a prepared edge profile between the parts.

### Welding of metals: principal considerations

It is safe to say that for low carbon steel—the world's number 1 structural metal—welding is *the* most important joining process. Nearly all cars built today have a welded unibody, to give one important example. For many other metals, including aluminum and stainless steels, it is also one of the key joining processes, but in these metals it is generally more difficult to make a high-quality weld than it is when using low carbon steel. To improve weldability, the precise composition of metal alloys is often adjusted: for instance, stainless steel AISI 316L is the welding-friendly version of the more common AISI 304. So selecting welding as a joining principle usually also requires an awareness of the importance of choosing an appropriate alloy (and filler).

---

**EXERCISE 12.3**

Look up the main difference in composition between AISI 304 and 316L. How does this explain the different weldability of these two stainless steels?

---

Prior to welding, the surfaces of metal parts must be thoroughly cleaned and their oxide layers removed. This can be done either manually or using machines or robots, and the process can involve mechanical means (e.g., grinding), chemical means, or both. The preparation can be time-consuming and increase welding cost considerably, but it is essential for weld quality.

Despite its popularity, welding of metals has several inevitable problems. It involves an intense localized input of heat while the rest remains cold, with the inevitable implications for residual stress or distortion (depending on the constraint imposed by the "welding jig," the clamping or fixing system used to keep the parts in place during welding). Because the peak temperature is close to or above the melting point, the positive effects of any prior work hardening or heat treatments are undone within the "heat-affected zone" around the weld (Figure 12.3). A modest loss of strength can be allowed for in design, but what is more critical is a loss of toughness and ductility. This is a particular risk with medium to high carbon and alloy steels—the good hardenability that enables the quench-and-temper heat treatment (Chapter 6) now becomes the opposite of weldability, as easy formation of (brittle) martensite is the least desired outcome. Careful alloy selection and control of welding parameters, with possible pre- and post-weld heat treatments, must be considered for steels other than low carbon steel. Heat-treatable aluminum alloys lose their high strength through welding, but weldable varieties have been developed that "naturally age" effectively—that is, they develop precipitation hardening at room temperature over periods of days or weeks, thereby allowing manufacturers to avoid the difficult and costly task of heat treating a complete product.

---

**EXERCISE 12.4**

The thermal conductivity of aluminum is approximately 2.8 times that of low carbon steel. What do you think this means for the efficiency of the welding process and the resulting residual thermal stresses?

---

Furthermore, as in casting, molten metal exposed to the air will react with any available oxygen during welding, and entrapment of oxides in the weld makes the material brittle and less resistant to fatigue or corrosion. This can, and usually is, prevented by welding in an inert atmosphere, using a shielding layer of inert gas, but this process adds to the cost. Finally, welding often generates toxic fumes and should therefore only be done in dedicated, well-ventilated areas (also because of the heat and intense light that can be released).

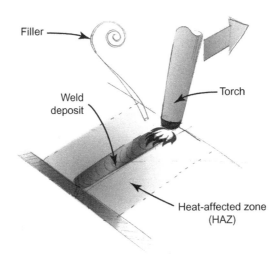

**FIGURE 12.3**

Schematic of the principle of (fusion) welding, with the heat-affected zone (HAZ).

## Welding of metals: main methods

For metals, there are literally dozens of welding methods in use today, and new ones are regularly being developed. Here are five of the more common methods (distinguished from one another by heat source), along with their main advantages and applications. *Many* more can be found in the literature.

In *oxy-acetylene gas welding*, the fuel gas acetylene ($C_2H_2$) is burned with oxygen to produce a flame that is used to melt the metal parts to be joined. A weld filler can be used, supplied in the form of wire or rod. The method is well-suited for continuous weld seams in thin low carbon steel products, welded in a single pass. The method can be automated, but it is essentially manual and hence slow. On the plus side, oxy-acetylene gas welding is flexible and requires only small product-specific investments (e.g., a welding jig), making it applicable to small production volumes.

*Tungsten-inert-gas ("TIG") welding* uses a tungsten electrode to set up a powerful electric arc between the electrode and the parts to be joined, thereby supplying heat (depending on the metal, either AC or DC current is used). Weld filler is added in the form of a wire, and a mixture of argon-helium usually serves as a shielding gas (Figure 12.4) to prevent oxidation. The method is particularly well suited to thin sheets and tubes of aluminum, magnesium, and titanium. For steels, the variant known as *metal-inert-gas ("MIG") welding* is used, the difference being that the filler wire itself acts as the electrode and is consumed. Both TIG and MIG are difficult to automate, and weld quality relies strongly on the welder's skill. Aluminum bicycle frames are nearly always made this way, leaving the patterned weld beads characteristic of this method (recall Figure 12.1).

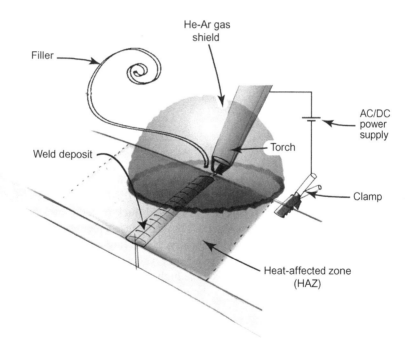

**FIGURE 12.4**

Tungsten-inert-gas arc welding.

*Resistance spot welding* is done by clamping sheet parts, usually low carbon steel, between two copper electrodes and discharging a strong burst of electric current. No weld filler is needed. Electrical heating is concentrated in the place of highest resistance—that is, the contact area, resulting in a lens-shaped weld "nugget" (Figure 12.5). Spot welding is fast (under one second for the actual welding, plus a few seconds to position the clamps), cheap, and gives little to no distortion. It can be done manually, with simple tools and jigs, or it can be fully automated using robots, making it applicable to a wide range of production volumes. Because of the clamping applied by the process itself, the gap between the parts prior to welding does not need to be closely toleranced, which is a distinct advantage. However, spot welds are not very strong (particularly with respect to fatigue), and the method is difficult with other metals, including aluminum. Automotive unibodies are the main application: a typical mid-sized car body contains thousands of spot welds.

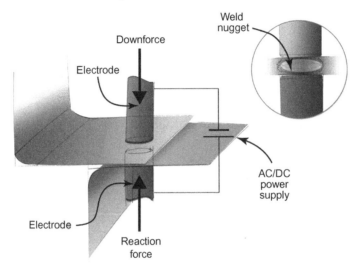

**FIGURE 12.5**

Spot welding set-up, including close-up of spot weld.

In *laser welding*, a powerful laser beam provides the heat for making a continuous weld, with speeds up to 10 m/min in thin sheets. It can also make spot welds (razor blades are an example). Unlike other methods for continuous welds, laser welding can weld thick material in a single pass weld, without filler, resulting in a narrow heat-affected zone. This is because a focused laser will vaporize the surface and the vapor absorbs the beam energy efficiently, generating a vapor "keyhole" through the metal, surrounded by a melt zone. For thin sheet and spot welds, the laser is de-focused and operates as a conventional heat source. Its capital cost is quite high, as it requires full automation, limiting its application to larger production volumes only—though "job shops" enable the cost to be distributed over a wide range of products at lower production runs, or even prototypes. As for metals, low-carbon steel is again the easiest to process. Laser welding of aluminum is more difficult (the metal reflects most of the light, and its high thermal conductivity transports the heat away from the welding zone quickly, making the process very inefficient) but can be done. The unibody of Audi's 1999 A2 supermini was a breakthrough application, and increasingly, the stiffening "stringers" inside aircraft fuselages are laser welded also (see also the discussion of laser cutting in Chapter 13).

*Friction stir welding* (FSW) is a relatively new solid-state welding method in which heating is by friction and intense plastic deformation, generated by a rapidly rotating shoulder with a protruding pin that is pushed along the joint line, again allowing manufacturers to avoid the need for filler (Figure 12.6). Because there is no need for melting and the hot deformation is applied instead, the resulting joints are strong and reliable. FSW can be surprisingly easy to set up, requiring little more than a standard CNC-operated milling machine for making small welds. But for larger products, large, dedicated stiff machines are required to maintain the downforce and traversing force on the tool with sufficient precision, strongly restricting the possible weld geometries. FSW was initially developed for aluminum to enable long continuous welding of extrusions to make structures such as ship decking and helicopter platforms. It is particularly attractive for the more reactive metals, such as titanium, as it avoids melting, but its use has also been extended to copper and steels. The 2012 Apple iMac ultra-thin desktop computer relies on this process, to give one recent example of its use.

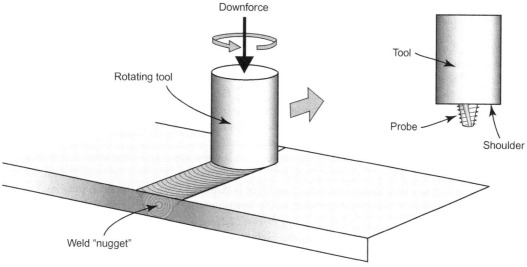

**FIGURE 12.6**

Schematic of friction stir welding (FSW) showing tool detail (top right).

---

**EXERCISE 12.5**

Summarize for yourself which of these five methods are applicable to carbon steels and which to wrought aluminum alloys.

---

## Welding of plastics: main methods

Many thermoplastics can be welded by heating beyond their glass temperature $T_g$ (for amorphous plastics) or melting temperature $T_m$ (for semi-crystalline plastics). Note that it is not possible to weld thermosets or elastomers—their permanent cross-linking means that they will not melt, but only degrade.

Compared to welding of metals, it is much easier to weld thermoplastics, as they require far lower temperatures, generate negligible distortion or changes to the material's microstructure, and produce little or no fumes. Still, it is not as popular as welding of metals, principally because there are often better joining options available, notably using form closures (e.g., "snap-fits"; see Section 12.7). Welding of plastics is therefore restricted to situations where the joint must be relatively strong, airtight, or both—domestic gas piping is an example—but even then, for many applications, adhesive bonding gives competition. Another difference with most metal welding processes is that because of the high viscosity of molten plastics and the need to inter-diffuse molecules to make a sound joint, the parts must always be pressed together firmly during joining and solidification. Note also that because of their high specific heat, plastics require more heat input than might be expected. Given their low joining temperature, and due to their low thermal diffusivity, it is not easy to put in this energy fast (recall that this was also a major issue in thermoplastic forming). In practice, this means welding of plastics is slower than you might expect.

Not surprisingly, there are fewer methods for welding of plastics than for welding of metals. But two processes are very common and important:

- *Ultrasonic welding* uses a piezo-electric element to generate ultrasound (typically around 15 to 70 kHz), which is projected into the plastic parts to be joined. The high frequency vibration causes frictional heating at the interface. Once heated, the parts are pressed together and left to cool down. Ultrasonic welding is cheap, relatively fast, and quite flexible in terms of joint geometry (although special joint shapes require dedicated piezo-elements). Its main downside is that the joint is not very strong; in this sense, the process is comparable to metal spot welding. USB connectors are among the many applications of ultrasonic welding of plastics.

- *Hot plate welding*, as the name suggests, applies heat via a hot metal plate that is then removed before the parts are pressed together (Figure 12.7). The hot plate, and thus the weld contour, is usually flat. The main benefit over ultrasonic welding is the weld strength; the main drawback is that this method is slower and more expensive, as the weld contour usually requires finishing. A typical application is the logo on automotive air bag covers (different logos can be used for the same cover part). This joint should be able to resist the high forces that occur during air bag deployment.

---

**EXERCISE 12.6**

Do you think it is possible to weld different kinds of plastics together (e.g., PS to LDPE)? If so, what would be the first requirement?

---

## 12.4 BRAZING AND SOLDERING

In brazing and soldering, a filler metal, usually supplied as wire, is first applied locally to the parts to be joined. The parts are then heated, thereby melting the filler and causing it to spread over the joint surface area through *capillary forces* (Figure 12.8) and then to solidify (the name *soldering* in fact comes from the Latin *solidare*, meaning "to make solid"). The parts that are joined—mostly metals, or

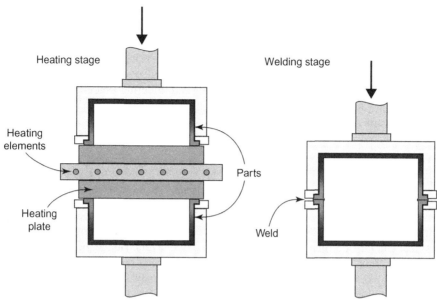

**FIGURE 12.7**

Hot plate welding setup. (For color version of this figure, the reader is referred to the online version of this chapter.)

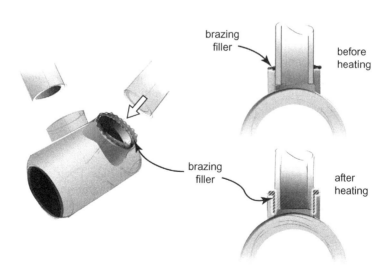

**FIGURE 12.8**

Brazing, as applied to a steel bicycle frame (sleeve joints).

sometimes ceramics with a metal coating applied to enable this type of joining—do *not* melt, and this is a key difference with fusion welding.

The difference between brazing and soldering is somewhat arbitrary—it is the temperature at which the process takes place, and thus the filler metals required: soldering is done below 450°C, with tin-bearing alloys; brazing is done well above this temperature, to join copper alloys such as brasses (hence the name *brazing*) but also other metals. For both brazing and soldering, eutectic filler compositions are preferred because they have a specific melting point (unlike most alloys that have a melting range) and a two-phase microstructure, giving greater strength.

---

**EXERCISE 12.7**

Eutectic compositions have another benefit in brazing and soldering. What is it? Hint: see Chapter 3.

---

Brazing and soldering both require the joint surfaces to be very clean, free from surface defects (e.g., burrs) that can impede filler flow, and free from oxide layers. Hence, before applying the filler, the process uses a "flux," a chemically active cleansing agent that removes oxide and remains on the surface to preventing re-oxidation. A second special "wetting agent" is usually applied for brazing, and often in soldering, to lower the surface tension between the filler and the parts to be joined, enhancing the capillary action. Because of these chemicals, brazing and soldering should be done within designated areas. Heat input may be modest compared to welding, but it is sufficient to generate harmful fumes that require ventilation. In high volume production, these issues require dedicated equipment, facilities, and training.

From the perspective of design, brazing and soldering mainly differ in joint strength, with the former being much stronger. Both can easily produce airtight joints. Furthermore, as the liquid filler tends to spread itself evenly, the joints can be very neat; for this reason, metal furniture and jewelry often involve brazing instead of welding. Another advantage compared to welding is that much less heat input is needed, reducing thermal distortion and residual stresses. And because the parts themselves do not melt, their microstructure is much less affected—and may even be enhanced. Less obvious, but important nevertheless, is that brazed and soldered joints can often be undone simply by applying sufficient heat, allowing repairs without undue damage to the parts.

## Brazing: principal considerations and methods

In principle, brazing can be used to join any combination of metals. Even certain ceramics can be brazed, provided they are first coated with copper or nickel. However, these applications are rare, and the process is more commonly applied to join metals, and then only like with like. The common low alloy steel bicycle frame is a prime example: the joints between the lugs and tubes are brazed (see Figure 12.8).

Provided they are designed and made well, brazed joints can be quite strong—even stronger than the parts themselves. Three main elements determine joint strength: (1) joint thickness, (2) joint area, and (3) the microstructure and strength of the filler, particularly at the interfaces between filler and components. Brazed joints are typically between 0.02 and 0.2 mm thick, depending on the size of the parts joined together. The thinner the joint, the higher its shear strength; tensile strength too is enhanced by keeping the thickness low, subject to a minimum value being used. Compared to joint thickness, the effect of joint area is more straightforward: the larger the area, the stronger the joint, explaining why lap and sleeve joints are *much* better than butt or tee joints. As for the microstructure, the strength of the filler determines when the filler will yield. However, interface toughness is usually more of an issue. Brazing involves some inter-diffusion between the liquid filler and the solid metal parts. Depending on the metals, intermetallic compounds may form, which can be stronger than the parts but tend to be more brittle. Under the right conditions, the transitions from base metal to filler can be well-graded and effectively invisible. Titanium brazed with pure tin shows this desirable result, although it does require additional heat treatment.

---

**EXERCISE 12.8**

Assume we braze two thin-walled high strength, low alloy steel tubes together with a sleeve joint. Calculate the overlap length that is needed to ensure that the joint is stronger than the tubes. For the joint's shear strength, assume a value of 20 MPa.

---

In industrial practice there are several brazing methods. They are differentiated by the heat source: torch or furnace (the most common), but also by induction, resistance, infrared, and so on.

*Torch brazing* is the common manual method, whereby a skilled worker manipulates a torch and a reel of filler wire to produce all manner of brazed objects. With the large number of variables, it is difficult to predict costs or cycle times, but to give some idea, an experienced worker can prepare and torch-braze all fourteen sleeve joints on a standard steel bicycle frame in 30 minutes.

In *furnace brazing*, the parts are prepared, mounted in a jig, loaded with filler wire in the right places, and the whole assembly is put in a furnace and heated until the filler liquefies and is drawn into the joint areas. On removal, the filler solidifies and the assembly is cooled and cleaned. Furnaces can be batch type or continuous, and they can have either an inert-gas atmosphere or (rarer) be kept under vacuum. Because the entire assembly is heated uniformly, complex shapes can be brazed. Furnace brazing requires little to no skill, beyond setting up and applying the filler correctly, but it requires more investment than torch brazing. For the same steel bicycle frame, this process would require less than 10 minutes of manual work.

## Soldering: principal consideration and methods

Sometimes called "gluing with metal," soldering is *the* main process for creating an electrically conductive joint between metal or metal-coated components. Printed circuit boards (PCBs) contain numerous soldered joints, and in this multi-billion-euro industry, automated soldering has reached very high levels of sophistication indeed. More humble are the soldered joints in, for example, a desk lamp, connecting the main wires to the lamp socket—but even there, production volumes have driven considerable refinement. Since the early 2000s, traditional, lead-based "solders" (i.e., the filler metals used for soldering) have made way for more environmentally friendly lead-free alternatives, such as tin-silver (96% to 4%) and tin-bismuth (42% to 58%) alloys. Soldered joints have low strength, and the failure of portable electronics is often due to a joint failing when the product was dropped.

Apart from electrically conductive joints, soldering is also applied on a large scale in heat exchangers, such as car radiators. Here, numerous flat, thin-walled (typically just 0.2 mm thick) aluminum tubes have to be joined at the top and bottom to collector plates, and although the required strength is low, these joints have to be watertight. For virtually all heat exchangers and radiators, wrought aluminum-manganese alloys ("3000 series") are used. This does not imply that aluminum is easy to solder: it is not, but given the huge market for heat exchangers, specific solutions have been developed. As a general guideline, copper and precious metals are easiest to solder, iron and nickel are more difficult, and aluminum and stainless steels even more so. Other metals (e.g., titanium, magnesium) are so difficult that in practice, first plating them—for example, with tin—is the better option. So coating steel sheet with tin not only gives the steel a corrosion-resistant surface suitable for food packaging, but it also facilitates soldering.

---

**EXERCISE 12.9**

What is the thickness of the tin coating on typical tinplate? How does this affect the recyclability of the steel?

---

As for soldering methods, there are plenty around, ranging from the simple manual use of a soldering iron for small production volumes to semi-automated furnace soldering for larger volumes. *Reflow soldering* deserves special mention. This process uses semi-solid solder pastes consisting of solder and flux, which can be pre-placed on flat parts via screen printing. It is used—fully automated and at a very large scale—to make printed circuit boards, with the solder paste joining the electronic components to the boards.

## 12.5 ADHESIVE BONDING

When asked what they think of adhesive bonding, many professional product designers say just one word: "avoid." This poor reputation is not entirely undeserved. Adhesive bonding does require careful preparation, even more than welding, brazing, and soldering, and it tends to be slow, and hence, unsuited to high volume production. Furthermore, its key strengths—airtight joints, no loss of material quality, and an absence of stress-raising discontinuities—are often not that important. Exceptions include aircraft structures, high-end racing bicycles, sports cars, and similar lightweight, low volume, high performance products, which often involve adhesive bonding. It is particularly important as the prime process for joining fiber composite materials, but new adhesives have also given a new lease on life to the most traditional structural material of all: wood. But the sub-principle deserves more attention for everyday products as well, if only because it can be combined with other joining processes to achieve valuable results. For instance, some carmakers now put adhesives in spot-welded joints to increase durability and reduce the squeaking and rattling that the body makes under stress, with no significant reduction of manufacturing speed.

Adhesive bonding is done as follows: after preparation, a suitable adhesive is applied to the joint area. Adhesion itself is a complex phenomenon involving both physical (e.g., materials inter-locking at a micro-level) and chemical (e.g., Van der Waals forces) aspects. The interplay of all of these factors is still not fully understood, but we do know the importance of proper preparation: removing oxide layers (if any), thorough cleaning, and adding wetting agents, all with careful control of cleanliness, humidity, temperature, and joint fit-up. After application, the adhesive requires time to set, often under controlled pressure and at increased temperature. All in all, the work is difficult to automate and therefore involves manual labor except in the highest of production volumes, which introduces problems of cost and quality control. Static joint strength and durability can be strongly reduced by seemingly small deviations from procedure, thus complicating joint quality inspection.

Joints made with adhesive bonding should be loaded in shear stress only, as the adhesives are relatively weak in tension—their tensile strength is just some 50 MPa at most and is often less. So joint geometries must be lap or sleeve joints, and although local spot joints are possible for non-demanding applications, continuous joints are recommended. Joint strength depends on the same factors as brazing: joint thickness and area, and adhesive strength and interface quality. Nearly all materials can be bonded, and the process is well-suited to joining dissimilar materials as well. Most adhesive joints are permanent.

Applying the adhesives onto the surfaces that will constitute the actual joint areas can be done in various ways, adding to the variety and complexity of adhesive bonding. For manual application, common solutions include tubes, spray guns, and tapes. The last solution is gaining popularity, with "mounting tapes" being used in an increasing range of products, from electronics to building applications. They can have the added benefit of filling any tolerance gaps between parts.

In principle, the whole process of preparation, application, and setting can be automated step by step, but for smaller production volumes, manual labor is the rule. Note that even adhesives that are relatively quick to harden, cure, or set times are still slow compared to other joining processes (except the well-known "superglues" that set nearly instantly, but these are very brittle). So it can be advantageous to combine adhesive bonding with spot welding or riveting: this way, a hybrid joint is made that has sufficient handling strength immediately and achieves its full strength later, without holding up assembly.

---

**EXERCISE 12.10**

Assume we make a lap joint between two aluminum sheets using adhesive bonding. At what overlap length will the joint become stronger than the sheets, assuming a joint shear strength of 10 MPa?

---

## Types of adhesives

There are literally hundreds of adhesives in use today, and finding the right one may require specialist advice, but they fall into four main categories.

*Drying adhesives* are dispersions of the active chemical adhesive compound (often polyvinyl acetate, PVA) in a solvent, such as water. Common adhesives in this category include those based on neoprene; the "animal glue" based on rendered collagen is another example. When the solvent evaporates, the adhesive hardens. Adhesives for wood are often of this type. To help the solvent to evaporate, the surface of the parts to be joined should be somewhat porous (as untreated wood is). Evaporation takes time, and joining with drying adhesives is therefore slow.

*Multi-component reactive adhesives* are mixtures of two different compounds that are inert while stored separately. Unlike the next category, mixing is done during application of the adhesive (not before), setting off the chemical reaction directly and releasing heat. Common combinations are poly-urethanes with polyesters, or polyurethanes with acrylic polymers. Recipes exist with and without solvents, meeting a wide variety of requirements.

*Single-component reactive adhesives* can consist of either a single compound or a mixture of two (so mixing is done before, not during, application); in both cases, the reaction requires a specific external trigger, which can be an ultraviolet (UV) light, heat, or moisture (which must be kept away while the adhesive is in storage). Common examples are epoxies (mixtures, triggered by temperature) and super-glues, i.e., cyanoacrylate-based adhesives (triggered by moisture).

*Hot melts* are thermoplastics that are heated until they liquefy, applied on the surface areas to be joined, then pressed and cooled. Many common plastics—such as LDPE, PP, PA, ethylene-vinyl acetate (EVA), polyesters, and (thermoplastic) polyurethanes—are available as hot melts, with specific additives for the application (e.g., "tackifiers"). Their main advantages over other types of adhesives are the higher production speeds they allow and the absence of chemical reactions, and hence possible fumes.

---

**EXERCISE 12.11**

Which two of these four types of adhesive can generally give permanent joints, and which two can also be semi-permanent?

---

## 12.6 MECHANICAL FASTENING

Mechanical fastening can be defined as joining two (or more) parts together by means of a separate component, relying on mechanical forces, friction, or form closure to make the joint. After the complexities and challenges involved in the previous three joining sub-principles, mechanical fastening comes across as relatively easy. It involves no heat or active chemicals, requires little to no preparation, and can be very fast. Not surprisingly, we often encounter it in the final assembly of products. Mechanical fastening is flexible regarding joint geometry and joint size (from tiny screws to huge bolts). Moreover, it can be used to join dissimilar materials, and many types of fasteners can be loosened repeatedly, making them ideal for parts requiring frequent maintenance or upgrading. Mechanical fasteners can even add functionality by enabling precise adjustments of part positions and joint strength. Finally, mechanical fasteners can be very strong, but they can obviously only be used for local joints, and making the joint airtight requires seals or gaskets.

There are numerous types of mechanical fasteners, such as bolts, screws, rivets, nails, staples, tie-wraps, cable clamps, and stitches. For reasons of brevity, only the first three are detailed next, with bolts receiving the bulk of the attention because of their high relevance to mechanical engineering.

### Threaded fasteners: bolts

Bolts may seem simple, but in reality the proper design for manufacturing of bolted joints requires careful study. Figure 12.9a shows two parts joined together by a nut and bolt with a "through hole." Notice how the bolt's shank is *not* in contact with the parts. Tightening the bolt puts the shank in tension, counteracted by compressive forces on the contact surfaces of the head and the nut, transferred

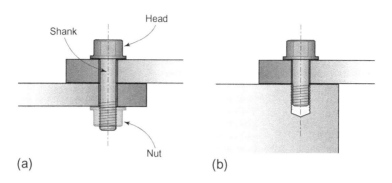

**FIGURE 12.9**

(a) A "through hole" (with nut and bolt). (b) A "blind hole" (with bolt only). (For color version of this figure, the reader is referred to the online version of this chapter.)

from the parts. Bolts can also transfer loads perpendicular to their axis—when the joint is loaded by tension in the plates, a "bearing load" is applied between each part and one side of the shank, and the load is carried by the bolt in shear (assisted to some extent by friction between the parts). The next most likely load to be transmitted through the bolt is bending, which redistributes the balance of tension/ compression in the bolt and parts. Note the danger of over-tightening—this takes the bolt closer to its yield stress than necessary, before an external load is applied, reducing the bolt's capacity for further loading. Bolted joints generally withstand fatigue loading well, especially if the parts joined together are much stiffer than the bolt itself. This is not only true for a through hole joint, but also for a "blind hole" joint (Figure 12.9b).

It is possible to make a bolt so that its shank does make contact with the parts: such "fitted bolts", also known as "close tolerance bolts," can carry significantly larger shear loads than normal bolts. They are, however, relatively expensive to make and are generally to be avoided in design and engineering.

Bolts themselves are usually made of steel, either low carbon steel for low strength bolts or hardened high carbon steel for more demanding applications. High strength bolts usually have designations such as "8.8" on their heads, indicating an 800 MPa tensile strength and a yield strength that is 0.8 times that value (i.e., 640 MPa). Carbon steel bolts are always coated to enhance corrosion resistance, with coating based on hexavalent chromium currently being phased out in favor of more environmentally friendly alternatives. For superior corrosion resistance, stainless steel bolts can be used (often with the designation "V2A"), but these are much more expensive. Aluminum bolts exist too, but their application is mainly limited to magnesium components (the magnesium BMW NG-6 engine block contains some 200 aluminum 6056-T6 bolts), and again at a price. Titanium bolts are even more expensive and used nearly exclusively in the aerospace industry. Whatever the material, all bolts are forged by a method called *cold heading*, which ensures a continuous grain structure for optimal strength and ductility (see Chapter 6). The method is reserved for large production volumes, which is partly why bolt sizes are standardized (in both metric and Imperial systems).

To ensure that the bolt's thread is strong enough to load the shank to its maximum strength, the thread depth and pitch scale with the bolt diameter. Furthermore, bolt heads often have a flange, dispensing with the need to have a separate load-spreading washer underneath and facilitating assembly. Most bolt heads have external or internal hexagon shape (the latter being known as an "Allen nut" or

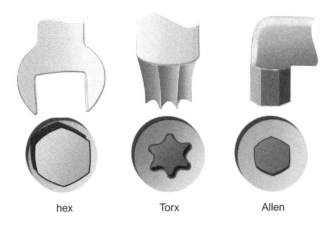

hex            Torx            Allen

**FIGURE 12.10**

Common bolt head designs.

hex socket, or an even a more complex shape, such as the Torx design) (Figure 12.10). Because of its superior formability, carbon steel offers the most options for such complicated heads.

Bolt tightening may also seem easy, but it comes with many challenges nevertheless, of which we mention the two main ones here. Tightening is done by turning the bolt with a certain torque, with 80% to 90% of this torque being carried away by friction in the thread and under the head. Because the precise amount of friction cannot be predicted well (threads can be damaged, increasing friction, or contain bits of oil, decreasing it), it follows that the exact clamping force may vary considerably also—and with it, the joint strength, which can easily be just 50% of the maximum possible value, even under controlled conditions. A second challenge is posed by *settling*. At each of the contact surfaces (see Figure 12.9), up to 0.05 mm of the shank elongation can be relaxed as the surface irregularities and coatings bed in. Because elastic bolt elongations are only fractions of a millimeter anyway, much of the clamping force is lost, particularly for short bolts used on rough surfaces. Re-tightening after settling is a solution but is awkward in high volume production.

---

**EXERCISE 12.12**

Assume we have an M6-bolt of strength class 8.8 with a 30 mm long shank. How much elastic elongation (in mm) can this bolt sustain? Will settling pose any problems? And how many turns of the bolt are needed to generate this elongation, assuming a regular thread pitch of 1 mm?

---

As for bolt tightening methods, we can distinguish between these four:

- *Torque limited.* In this common method, either a manual torque wrench or an automated setup tightens the bolt to a pre-set torque value. This method is cheap, but its downside is the high variation in clamping force mentioned previously.
- *Torque plus angle.* This method tightens the bolt to a pre-set torque value and then tightens it over an additional angle to ensure a certain minimum elongation (e.g., if the thread pitch is 1 mm,

90 degrees extra means 0.25 mm extra elongation). It is more complex and costly to set up but reduces the variation in clamping force.

- *Tighten to yield.* By measuring torque and angle simultaneously, it is possible to tighten the bolt until its yield strength is reached, at which point the angle begins to increase faster than torque. This gives lower clamp force variation but does not leave capacity for carrying additional loads through the joint.
- *Ultrasonic tightening.* In this novel method, developed under the name Nedsonic, the bolt elongation is measured continuously using ultrasound, bypassing the problem of friction and delivering minimal clamp force variation with a dedicated but relatively simple hand tool. Its main drawback is that it currently only works for bolt diameters sized "M8" and up.

## Threaded fasteners: screws

Both screws and bolts are "threaded fasteners," but it is not easy to define the precise differences between them; indeed, the various definitions in the literature and in industry are often contradictory. Here, we define the first key difference as follows: a screw *does not* have a free shank and hence can transmit shear forces via direct contact (i.e., form closure and friction), whereas a bolt *does* have a free shank and hence can transmit shear forces only via clamping force and friction—fitted bolts obviously being an exception to this rule. The second key difference is that unlike bolts, screws form their own thread when they are screwed into a pre-formed hole (either by deformation or by cutting) and are kept in place through friction. All screws are "self-tapping"; some "self-cutting screws" also drill their own holes, but usually a hole is required to create the space that will be occupied by the screw once inserted. Consequently, screws are much less suited for repeated loosening than bolts.

A third major difference is that in screws, thread design is strongly optimized toward the materials they join, instead of standardized with a 60-degree thread angle, as it is for bolts. So screws for plastics have different thread designs than those for wood or metals. Indeed, screw threads are often specific for certain grades of plastics: for instance, glass-filled ABS requires a different screw thread geometry than unfilled ABS. Still, especially in plastics, the thread is basically a crack, and the actual strength of screwed joints is easily overestimated (particularly long-term strength, when fatigue or creep may come into play).

Numerous screw head designs exist, either with the head sticking out of the joint or with it being "counter-sunk", flush with the surface. Figure 12.11 presents three common varieties. Screws are usually made from low carbon steel (by cold heading, much like how bolts are made), but screws from stainless steel, aluminum, brass, or other materials exist as well. As screws typically join materials that are weaker than the screw material itself, there usually is no need for high strength screws. Exceptions are self-tapping screws that must be able to drill their own holes into strong materials; these are made of high carbon steel and usually have induction-hardened cutting tips.

---

**EXERCISE 12.13**

Find illustrations of both self-tapping and self-cutting screws. What is the key difference?

---

Screws are generally put in using manual labor, with simple, low-cost tools. For very large production volumes, the process can be automated, but this requires dedicated and costly investments for

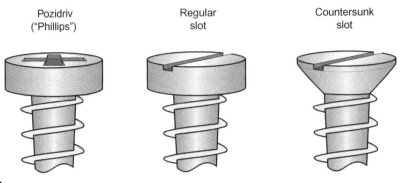

**FIGURE 12.11**

Three common screw head designs: Pozidriv or "Phillips", regular slot, and countersunk slot.

screw supply and placement. Whatever the situation, the best way to keep assembly costs down is not only to minimize the number of screws but also to minimize the number of different *types* of screws—indeed, such standardization is beneficial for any kind of mechanical fastener and is employed by countless manufacturing companies.

## Rivets

The third type of mechanical fastener discussed here is the rivet, shown in various forms in Figure 12.12. Rivets keep the Eiffel Tower together, as well as heavy-duty truck frames, but also modern airplanes: the Airbus 380 superjumbo and Boeing 787 dreamliner contain hundreds of thousands of rivets.

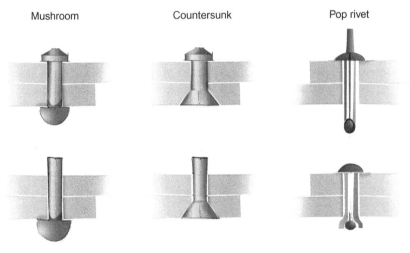

**FIGURE 12.12**

"Mushroom," countersunk, and pop rivets, before placement (upper row) and after (lower row). In each case, the tool operates from above.

Many consumer products also rely on rivets (e.g., baby strollers). The process works as follows: A metal rivet is placed through a hole in the two parts to be joined, with the rivet shank being some 10% thinner than the hole diameter. Next, the rivet is compressed and plastically deformed until it fills the hole and is deformed with a second "head" on the rear side of the joint (see Figure 12.12). Countersunk rivets exist as well: these lie flush with the joint surface on one side. Unlike bolts and screws, rivets are permanent, but they usually are cheaper. So rivets are essentially form closures, with the added bonus of leaving compressive residual stresses around the hole, improving fatigue strength. Notice that metal rivets undergo considerable strain hardening during the process. Without this increase in rivet strength, the joint would not be strong enough.

The first two rivet types shown in Figure 12.12 require access on both sides of the joint for application. In practice, this can be done with a heavy clamp reaching around the joint, but this restricts joint geometry. The third type of rivet does not have this disadvantage: this "pop rivet" requires one-sided access only. Most pop rivets are not very strong—in single shear loading the joint will twist and the rivet pops out. There are exceptions that can even replace heavy duty M12-sized bolts at equal strength; however, such rivets are expensive. "Self-piercing rivets", which create the rivet hole and joint in one operation, are increasingly popular in automotive manufacture, the higher rivet cost being offset by simpler, faster joining times. Riveting, like bolt fastening, nearly always involves manual labor using dedicated tools. Automation is rare and limited to specific cases (e.g., aircraft structures). Finally, like bolts and screws, rivets can also be used to create a specific look and feel.

---

**EXERCISE 12.14**

In practice, both the rivet itself (length, diameter) and the hole in which it is placed have tolerances. How do these affect the eventual joint strength?

---

## 12.7 JOINING USING FORM CLOSURES

The fifth and final sub-principle is joining using form closures. It is potentially the neatest of them all because it requires no heat or chemicals nor any additional materials or components. Form closures can be strong, invisible, join dissimilar materials, allow repeated release, and can be very rapidly made. A varied range of methods can be split into two categories: form closures based on *elastic* deformation of parts and those based on *plastic* deformation.

### Elastic form closures

As the name implies, making these form closures avoids yield of the material. Consequently, they can, in principle, be released repeatedly. Elastic form closures are best done with strong, low-stiffness materials—that is, those that allow relatively large elastic strains (principally plastics and, to a lesser extent, aluminum). Two common methods are snap fits and clamp fits.

### Snap fits

Countless plastic products contain these joints, and Figure 12.13 shows the general principle. The force required for closing is influenced by the magnitude of the "entering tip angle" $\alpha$ (as well as snap fit

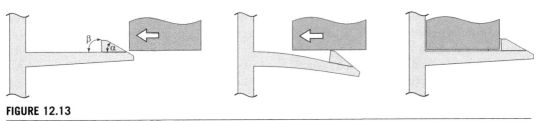

**FIGURE 12.13**

Schematic of flexible snap fit.

length, material, etc.), whereas the "withdrawal tip angle" β determines if the joint can be loosened again without special tools. The battery cover of your TV remote likely contains this type of joint. Placement is generally done manually, either by hand or with simple tools, but as this can be done fast, snap fits are suitable to large production volumes. As they involve undercuts, the molds get more expensive, but plastic snap fits repay this extra cost many times over through the much easier assembly. In aluminum extrusions, snap fits can often be integrated into the profile cross-section at low extra cost, but as the aluminum surfaces tend to stick more readily than plastics, loosening such joints can be difficult.

### Clamp fits

Given the right geometry and tolerances, we can simply press or slide one part or component into another with an "interference fit," meaning that the parts hold together by the process leaving an elastic compressive stress (with friction) across the interface. Examples include dovetail joints in wood, metal press-in bearings in plastic housings, and so on. When both parts are made of metal, the required force may get high, but if time is not too much of an issue, we can often heat the outer part to expand it, slide it in place, and let it cool down. On some bicycles, the joint between the pedal crank and the main pedal axle is made this way. Clamp fits are semi-permanent, as they can usually be loosened with some effort.

---

**EXERCISE 12.15**

Determine the maximum elastic deformation (in compression or tension) that standard ABS can withstand. How does this compare to a grade with 30% glass fibers? What does this suggest about the suitability of snap fits in both situations?

---

Note that in many products, parts can simply be joined by enclosing them on all sides by other parts, sometimes requiring zero elastic deformation (for an example, consult Chapter 2, Figure 2.5). Alternatively, given the right shape, we can hook one part around another, leaving it free only to rotate, then take away the last degree of freedom by a single snap fit (or some other kind of joint, such as a "grub screw" that presses against a flat on the inner part).

## Plastic form closures

These types of joints are only available for parts made from materials that have plastic strains (i.e., metals or heated plastics) and result in permanent joints. Plastic strains can be large, and it therefore need not be a problem if the parts to be joined have relatively wide tolerances. There are three common methods: seaming, clinching, and staking.

In *seaming*, two metal sheets are folded over each other to create a continuous joint (Figure 12.14). This is essentially done in three steps: first bending the sheets to create "hooks," then hooking the two parts into each other, and finally rolling to consolidate the joint. For sheets with thickness $t$, the geometry is constrained by the minimum bending radius $MBR$, which in a first approximation is equal to $MBR = t/(2\varepsilon_{max})$, with $\varepsilon_{max}$ being the strain-to-failure (see Chapter 4 for details). Seaming can be done with simple tools, is relatively quick, and is suitable for a wide range of production volumes. Cans for canned foods are among the numerous applications.

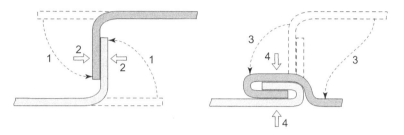

**FIGURE 12.14**

Seamed joint between two metal sheets. (For color version of this figure, the reader is referred to the online version of this chapter.)

*Clinching* is somewhat comparable to seaming and entails making local rather than continuous joints. It involves interlocking two metal sheets into each other at one point (Figure 12.15). This requires considerable ductility, making this technique primarily suitable to low carbon and low alloy steels, and less suitable to, for example, aluminum. The method is increasingly replacing other joining methods, such as riveting (a key benefit being that no separate fastener needs to be supplied). Clinching can be integrated into metal part forming operations, making it very cost effective in large production volumes.

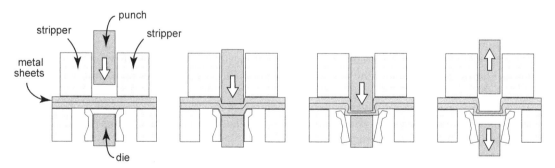

**FIGURE 12.15**

Schematic of clinching. (For color version of this figure, the reader is referred to the online version of this chapter.)

*Staking* can be used to join a metal strip to the inside of a plastic housing. The strip, containing one or more holes, is first located over plastic "stakes" protruding from the housing's inner surface and is molded as one piece with it; next, the stakes are heated and deformed to lock the strip in place

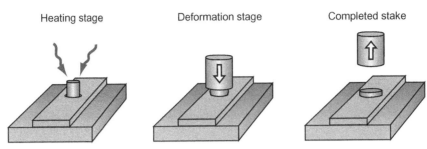

Heating stage          Deformation stage          Completed stake

**FIGURE 12.16**

Schematic of staking. (For color version of this figure, the reader is referred to the online version of this chapter.)

(Figure 12.16). If insert molding is not suitable, staking can be a good alternative. It is used in larger production volumes only—for instance, to join printed circuit boards to electronics casings.

---

**EXERCISE 12.16**

Return to your sample products (Exercise 12.1). What can you now say about the types of joining that were involved? For instance, can you pinpoint the method now as well?

---

## 12.8 ASSEMBLY: BASIC LAYOUTS AND CONSIDERATIONS

To close this chapter, we make a number of observations that designers should be aware of relating to *assembly*—that is, the sequential joining (and finishing) operations in a production environment.

The first observation is that some of the five joining sub-principles require dedicated environments and separate workspaces. This is true for welding, brazing, and soldering, as well as for adhesive bonding; mechanical fastening and joining using form closures generally do not need such separate facilities. For this reason, whether it is welded, brazed, or adhesively bonded, a product such as a bicycle frame (to return to the example from the introduction) is always joined "offline" in a separate workstation (even locations), to be fed into the sequence of bicycle assembly operations as a single component, often undergoing some kind of finishing operation (e.g., painting) before the other components are assembled to it. The take-home lesson is that joining provides *structure* to the entire assembly process, and vice versa, that assembly considerations inform choices for joining processes. Notice that the separation of joining operations also provides the opportunity for the work to be inspected at certain intermittent points, allowing reproducible results.

The second observation is that we can organize assembly in two essentially different ways, known as *line assembly* and *cell (or group) assembly*. In line assembly, the product moves along a number of stations, with a single operation taking place at each station. In group assembly, all operations are done at a single station. Line assembly tends to be faster and, for larger production volumes, cheaper. However, it is more difficult to design for, as each assembly operation has to take place in the same amount of time. Also, if done by humans instead of robots, it tends to give less job satisfaction: every worker then becomes merely one cog in the machine. In practice, combinations of these two ways are common, as in the bicycle example: cell assembly of the frame and line assembly of the complete bicycle.

The third observation is that assembly can be entirely manual, performed entirely by robots, or any combination in between. Particularly in industrialized countries, where wages are high, most assembly work is now done by robots, but humans can rarely be replaced completely, being endlessly more flexible than robots in terms of their mobility, ability to access difficult locations, and potential to improvise and modify in real time rather than repeat a pre-programmed process. And note that for an increasing number of products, assembly is done not by paid workers but by the end user: the world-famous IKEA bookshelf Billy (with a total production volume of more than 40 million units) is a prime example of such flat-pack, ready-to-assemble products. As a general rule in design for assembly, always consider who is doing the assembly and consider that person's level of skill and motivation.

The final observation is that, as mentioned in the introduction, many products are made in different varieties, usually in the same factory and often on the same assembly line (or cell). Bicycle frames, for instance, come in a range of sizes and colors and can be assembled into finished products with different specifications—standard bikes, "deluxe" bikes with suspension and extra gears, and so on—all on a single assembly line. The same happens for cars, electronics, furniture, and countless other products. At the start of assembly, we have the individual components that can still be used to make any member of the "product family," whereas at the end we have specific products. So at some point in between, the product starts to become customized (i.e., made for a specific client). From the manufacturing perspective, the challenge is to postpone this point in the assembly sequence as long as possible, thereby facilitating production and shortening delivery times for made-to-order products. Customization of assembly is primarily a logistical challenge that involves managing information flows with proper planning and advanced information technology systems to keep the supply of correct parts going with the minimum of (expensive) stockpiling onsite. For cars and other high-value products, such customization is now common—take any opportunity you can to visit a large-scale production line: it is a real eye-opener to see, for example, car bodies being robotically welded together in seconds, and to appreciate the remarkable technical achievement of cars rolling off the assembly line at roughly one per minute, with every vehicle being different in some respect.

---

**EXERCISE 12.17**

Assembly can be fully manual and fully robotized, and it can take place in either a line or a cell layout. Ignoring intermediate forms, this gives four possible combinations. Which one is "easiest" to design for, and why?

---

## 12.9 **Further Reading**

Joining and assembly is a broad and diverse field, so this chapter has necessarily presented a selection of the main methods and solutions. Among the many omissions are practical joining solutions such as press-in nuts (commonly used to provide attachment points to sheet metal parts for bolts) and all sorts of springs and clips used to keep parts in place—if you have done some product disassembly studies, you will no doubt have come across such handy solutions. But remember these main points: (1) the key attributes listed in the manufacturing triangle have been discussed and expanded in the context of joining; (2) joining is absolutely not an "add on" to be addressed only at the end of the design process, but must be considered early on (typically already during concept design) to allow choices for materials and shaping processes that are compatible with joining and thus meet the target functional, quality, and cost requirements for the product as a whole; and (3) joining processes have clear physical characteristics that explain, along with practical considerations and the state-of-the-art, what they can and cannot do, exactly as for shaping processes (Chapters 3 through 11). With this background, you are equipped to ask the right questions, whichever joining process, assembly operation, or design problem you encounter. In addition, there are many sources that you can access to explore the field further, some of which are listed here.

Scientific journals on joining and assembly tend to be organized much like the industry itself (i.e., by sub-principle). Examples are *Science and Technology of Welding and Joining*, which despite its title is firmly focused on welding (of metals, much less of plastics), *International Journal of Adhesion and Adhesives*, which focuses on adhesive bonding, and *Assembly Automation*, which covers assembly in its various forms. From a design perspective, these sources are usually suitable for background reading only, looking (as scientific papers mostly do) deeply into a very small area, though journals on manufacturing technology often take a more design-oriented approach.

Several suppliers of joining technology offer interesting information and design rules on their websites. Our product design specialists recommended, for example, www.fabory.com and www.ejot.de. Designers who are looking for news on joining and assembly will find the industry publication *Assembly* quite accessible (visit www.assemblymag.com). The organization behind this journal also organizes industry fairs. A different way to build up knowledge of joining and assembly is to perform product disassembly studies, as shown in Chapter 2. For inspiration, take a look at www.ifixit.com.

Bonenberger, P.R., 2005. The First Snap-Fit Handbook—Creating and Managing Attachments for Plastic Parts, second ed. Carl Hanser Verlag, Berlin. (*Well recommended by designers, this book is ideal for digging deeper into the number 1 joint for plastic parts, the snap-fit.*)

Boothroyd, G., Dewhurst, P., Knight, W.A., 2010. Product Design for Manufacture and Assembly, third ed. CRC Press. (*Still the benchmark for "DFMA" and highly recommended for a fresh design perspective. See also www.dfma.com.*)

Ferjutz, K., Davis, J.R. (Eds.), 1993. ASM Handbook Volume 6: Welding, Brazing and Soldering. ASM International, Materials Park, Ohio, United States. (*Focused more on engineering than design, much like all other ASM Handbooks, this is still a key reference for these joining processes.*)

Houldcroft, R., 1990. Which Process? Abington Publishing, Cambridge, UK. (*The coverage is not as broad as the title suggests, being largely specific to welding of carbon steels, but the detail in this area is good and highlights the coupling between part thickness, joint geometry, and process.*)

Petrie, E.M., 2006. Handbook of Adhesives and Sealants, second ed. McGraw-Hill, New York. (*There are many handbooks on this topic, many of them good. This one is excellent and carries our recommendation for those wanting to read further.*)

Womack, J.P., Jones, D.T., Roos, D., 1990. The Machine That Changed the World. Free Press, New York. (*If you want to know more about lean manufacturing and about how DFMA can drive design, this book is a great place to start reading.*)

# None of the Above

## CHAPTER OUTLINE

## 13.1 INTRODUCTION

The previous ten chapters have covered ten common, important manufacturing principles, along with the main methods we can place under them. This is an excellent start, but there are obviously many more. To keep this book manageable we stop the in-depth coverage here, but we will acquaint you briefly with some specific interesting processes, which are as follows (along with the materials they are best suited for):

- Semi-solid processing: thixomolding (magnesium)
- Powder methods: sintering (metals, ceramics)
- Laser cutting (mainly metals, but also many other materials)
- Rotational molding (thermoplastics)
- Extrusion (thermoplastics)
- Compression molding (mainly thermosets, but also several other materials)
- Press-blow molding (glass)
- Slip casting (ceramics)
- Surface heat treatments (metals)
- Coating processes (all materials)

There is no single rationale behind choosing these processes. Some came out of the initial selection made by professional designers and engineers that led to Chapters 3 through 12. This is the case for laser cutting or rotational molding, processes that, although certainly not in the same league as injection molding or machining, still deserve a mention. Others illustrate what manufacturing can do for materials other than metals or plastics (such as press-blow molding of glass and slip casting of ceramics). Finally, the last two on the list are included to invite you to think beyond shaping and joining, toward finishing operations. The format used in this chapter is somewhat different, briefly addressing (where relevant) the basis and history of each process, then the challenges, costs, and any competing processes, allowing you to better estimate their potential. No questions are included, because for a first encounter, questions would not be meaningful.

However many processes we cover, you will always encounter more. So the question is, how can you obtain in-depth understanding of *any* unfamiliar process? To address this question, we end the chapter with notes on how to collect useful information and build up solid insight about the manufacturing processes you may encounter in your professional career. For this, you'll need to cultivate the following skills: knowing what sources are available to you, with the ability to judge if a source is any good; and, knowing what you can find on the Internet—and whether you can trust the information.

## 13.2 SEMI-SOLID PROCESSING: THIXOMOLDING
### At-a-glance
Semi-solid processing (SSP) is a hybrid between casting and forging. From a design perspective, its strengths are easily summarized: SSP combines the form freedom of casting with the material quality of forging, offering rapid production speed and excellent control over tolerances. Its main disadvantages, apart from being relatively young and unknown, are the high investment costs, the limited part size, and the very small number of materials to which it can be applied. Of the various SSP methods, *thixomolding* is the most common. Several modern cell phone and camera casings are manufactured with this method, as is the frame of the GoCycle folding bicycle (Figure 13.1). By far the most popular thixomolding alloy is AZ91D (magnesium with 9% aluminum and 1% zinc). Production volumes are high (anywhere up from 100,000 units/year), size is limited (the GoCycle frame currently contains the largest thixomolded parts in the world), but tolerances can be very close and (for small products) wall thicknesses can be as low as 0.4 mm.

The secret of thixomolding is to process the material as a solid-liquid mixture. At the right temperatures and at high shear strain-rates, certain metals lose much of the shear strength they have at low strain-rates (Figure 13.2). In this state, the metals appear solid and can carry loads, but can be spread out with a knife, like butter. This peculiar phenomenon is known as *thixotropic behavior* and you can observe it in modern thixotropic "non-drip" paint, as well as in ketchup! In metals, its microstructural basis is twofold: (1) the metal must have spherical grains, and (2) it must have the right mix of solid and liquid phases (30% to 65% solid). This combination allows shaping of parts using relatively modest shear forces, with the solid-liquid "slurry" that has a sufficiently high viscosity to prevent turbulent mold filling, giving superior material quality. A third, more practical condition is that the behavior must persist over a sufficiently wide temperature window to allow processing into complex shapes. This implies that the metal must have a wide melting range—that is, a large difference between liquidus and solidus temperatures (the opposite of what we usually seek in a casting alloy). Also, the slurry should not damage the (steel) molds, which rules out aluminum. In practice, this means that only certain magnesium alloys are suitable.

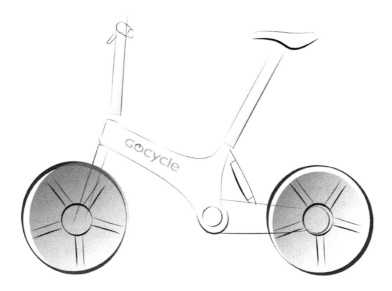

**FIGURE 13.1**

GoCycle. (For color version of this figure, the reader is referred to the online version of this chapter.)

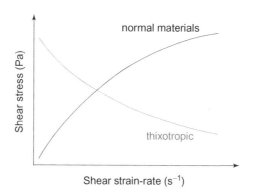

**FIGURE 13.2**

Shear stress against shear strain-rate, for thixotropic and normal materials. (For color version of this figure, the reader is referred to the online version of this chapter.)

## Process history

Thixotropic behavior in metals was discovered in 1971 by David Spencer while a PhD student at MIT with metals casting specialist Merton Flemings (known, among other things, for the "Flemings equation"). Though they quickly realized its manufacturing potential, it still took 25 years for its first commercial realization (as is common for material innovations). In the mid-1990s, two Italian companies applied thixomolding to car fuel rails, thus realizing the first successful application. In parallel, the Japanese company JSW developed the manufacturing equipment and currently leads the world market, with press capacities up to 800,000 tons. More recently, thixomolding moved into the world of electronics casings, centered in Malaysia and Japan.

## Challenges and costs

Thixomolding is anything but easy. Even for AZ91, the process temperature window is so narrow that injection has to happen *extremely* fast (in microseconds); otherwise, the material cools down too much through mold contact. Combined with the high viscosity, this leads to injection pressures way beyond those used in high-pressure die casting. So even small products require huge machines—with very advanced process control. Not for nothing has the U.S.-based company Husky, a giant in injection molding machinery, canceled its thixomolding developments, leaving JSW as the sole supplier of equipment. Machine costs are therefore relatively high, and mold costs very high. A second challenge is de-molding, which requires special mold release agents that need to be thoroughly removed after a production run: essentially, the entire mold (which is a complex machine in itself) must be disassembled, cleaned, and reassembled. For very large production volumes this is not a problem, but it is otherwise prohibitively costly. Recycling of production scrap (let alone post-consumer scrap) is another challenge, as it cannot be simply ground up and put back into the machine but requires specific treatment first. So thixomolding is strictly for the largest markets, automotive and electronics, with limited part sizes. At least the base material price is not an issue: per kilo, magnesium can be as cheap as aluminum.

## Competing processes

Thixomolding of magnesium (also known as "magnesium injection molding") also has stiff competition, notably from high-pressure die casting, which offers more form freedom but at reduced material quality, in particular lower strength and ductility. Integral machining of wrought magnesium is a second competitor, offering even more form freedom (undercuts, different thicknesses, etc.) with better quality and strength quality but at higher cost. Contrary to common perceptions, the fire hazard associated with processing magnesium can be kept well under control. A third competitor is injection molding of reinforced thermoplastics; for example, "polyarylamides" with 40% to 50% of glass fiber filling, marketed by Solvay under the trade name Ixef, offer similar strength and stiffness as AZ91D, with excellent surface finish and tolerance control—and at competitive prices. With all of this competition, it is clear that SSP is not a process you should choose lightly. Successful applications depend on thorough design and engineering as well as making careful trade-offs between the various aspects of the manufacturing triangle.

## 13.3 POWDER METHODS: SINTERING

### At-a-glance

There are several shaping processes that start with powder (usually metals, but also most ceramics), collectively referred to as "powder methods" or (for metals) "powder metallurgy", all involving some combination of high pressures and temperatures. A common member of this family is sintering. In this manufacturing principle, a precisely measured amount of metal powder is put into a strong die and compressed into the required shape. Next, this shape (known as the "green form") is ejected from the die, placed in an oven, and heated to around 80% of the powder's melting temperature, usually under a corrosion-inhibiting atmosphere. This lets the powder fuse together by *diffusion* of atoms, mainly along grain boundaries, to fill the pores. The same principle applies to ceramic powders: most technical ceramics are shaped this way. Plastics can also be sintered, although this is done only for specific applications, as there is little need to avoid melting. For metals, part sizes range from grams to kilos and production volumes range from thousands to millions per year (the heating step is slow, but ovens can process large batches at once). It can produce moderate shape complexity with close tolerances, and it is "near-net shape," indicating that no subsequent machining or finishing of parts is necessary (or practically viable, for ceramics). The range of powder methods is extended by the continued application of hydrostatic pressure during the sintering step, giving "hot isostatic pressing" (HIPing).

A key benefit of sintering is its suitability to metals that are difficult, if not impossible, to process by other means. Examples include tungsten and molybdenum, which have such high melting points that casting is challenging (with few, expensive, mold materials available) and such high hardness that machining is very slow. Sintering can also handle *mixtures* of metals that cannot be cast without segregation of the individual components, so the process allows manufacturers to "tune" materials to have the right trade-offs among strength, conductivity, thermal expansion, and so on. Various types of steel can be shaped this way, offering a welcome alternative to steel machining (slow) or casting (difficult). Note that "sintering steels" are very different in composition and material behavior from carbon or low alloy steels. A third distinguishing feature of the process is that, depending on process conditions, not all of the space between the powder grains is removed, leaving (controlled) porosity behind. This is ideal for bearing liners and other load-bearing, lubricated components (in which lubricant is retained by the surface porosity). Sintering applications exploiting these benefits are common (with 70% in automotive), though most are invisible to users: engine components, bearings, and bushes, but also permanent magnets, surgical implants, cell phone vibrator weights, and so on. One visible exception is the osmium-based tip of a fountain pen.

## Process history

Sintering, like casting, goes back to almost prehistoric ages, but one of the first modern uses came around 1900 in the form of sintered tungsten filaments for incandescent light bulbs. Since then it has seen steady growth to a point where the process is now fully mature. Future prospects include the increasing use of sintering for aircraft structures, replacing sheet metal assemblies of many separate parts by one-piece components.

## Challenges and costs

Form freedom in sintering is limited in a peculiar way that becomes apparent once we take a closer look at the compacting step (Figure 13.3). As the pressures during this step range from 700 (for aluminum) to 8,000 bar (for iron), the forces are huge, even on small components. The component should therefore ideally have two perpendicular planes, so that all of the compacting force is carried by the material. And, of course, it should have no undercuts, as is usual for principles and methods involving permanent dies. However, if there are differences in section thickness between the two sides of the press, then if a single die is used, traveling a fixed distance during compaction, there will be differences in the degree

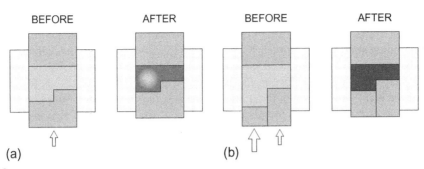

**FIGURE 13.3**

Powder compaction of a stepped product using (a) single-action press and (b) double-action press (giving more homogeneous compaction). (For color version of this figure, the reader is referred to the online version of this chapter.)

to which the powder is compacted (see Figure 13.3), which remain after sintering. This limits form freedom, especially if the part must have a certain final degree of porosity. Multi-action presses offer a way around this limitation, but they are somewhat more expensive.

The main part-specific investments for sintering are those required for the high strength tool steel dies. These costs are prohibitively high for small production runs. At high production volumes, the material price and general overheads become the dominant factors in the cost per part. A benefit of sintering is that it has zero waste, other than any rejects due to quality problems.

## Competing processes

The unique strengths of sintering ensure that, for many applications, there is no significant competition. Exceptions involve components sintered from low- to medium-strength steels, for which extruded and machined aluminum profiles provide an alternative, provided that (1) the production volume is not too large, (2) a yield strength of around 250 MPa is sufficient, and (3) application temperatures remain under 120°C. A second competitor for sintered steel is nodular cast iron, again combined with machining. Note that direct replacement of an existing sintered component with an identical geometry in aluminum or cast iron (or any other material) will rarely be optimal, but full redesign is needed to exploit the shaping benefits of the competing processes.

## 13.4 LASER CUTTING
### At-a-glance

Since their invention in the late 1950s, lasers have revolutionized countless products and industries, from data storage and retrieval (CDs, DVDs) to surgery, from holograms to printers, and from simple pointers used in presentations to equipment for measuring the distance to the moon. If you are looking for examples of disruptive technologies, then lasers are as good as they get! Not surprisingly, they have also affected the manufacturing industry. We already came across laser *welding* in Chapter 12; here, we introduce laser *cutting* as a separate manufacturing principle.

Modern $CO_2$ laser cutters routinely cut through many types of materials by focusing the laser beam on a narrow cut and vaporizing the material. With continuous output power levels of 10 kW or more, material thicknesses can easily go up to 10 mm, and cutting speeds are between 0.1 and 10 cm/sec, obviously depending on the precise power, material, and thickness. But speed and power are only part of the story. Laser cuts are very neat, smooth, and thin, as narrow as 0.1 mm for thin-gauge materials, especially if the smaller Nd:YAG lasers are used. Moreover, as lasers are comparatively light and small, they can easily be mounted onto *XY*-tables (Figure 13.4), which enables extremely complex contours to be cut with high accuracy and with *zero* product-specific investments! All that is needed is the CAD-CAM file, and laser cutting is therefore gaining ground as a rapid prototyping technology. In series production, it is also steadily replacing mechanical cutting processes because of its higher accuracy (no tools that wear out, no burrs that complicate assembly or injure users, no warping due to clamping) and easier set-up. For instance, the blanking step that precedes sheet metal forming is now done more and more using lasers. One specific high-volume application is *tailored blanking*. In this process, a laser first cuts two sheets with different thicknesses into the required shapes, and the same laser then welds these two blanks together to yield a product (e.g., an automotive door panel) with two different thicknesses.

If we put the laser not on an *XY*-table but on a robotic arm with sufficient degrees of freedom, laser cutting need no longer be limited to flat sheets but can also be used to accurately cut in three dimensions (3D). One application involves finishing the ends of bent metal tubes and profiles; together with modern CNC bending equipment and laser welding, this has the potential to revolutionize space frame design and manufacture. British automotive engineer Gordon Murray already uses this combination of manufacturing processes for his remarkable T.25 city car. It is challenging to do and will therefore likely remain a niche application, but a very interesting one nevertheless.

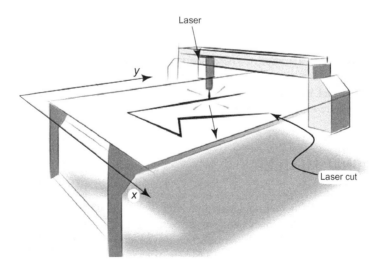

**FIGURE 13.4**

Laser cutting with *XY* table.

## History and outlook

The principle of the laser, or "light amplification by stimulated emission of radiation," dates back to a theoretical paper by Albert Einstein in 1917, but it would take some 40 years before a working laser was first built, producing visible light (Bell Labs, 1958). Shortly afterward, the first gas lasers were developed, which are lasers that operate by discharging an electric current through a gas. One of these is the $CO_2$ laser, using carbon dioxide and producing near-infrared light. Invented in 1964, this is still one of the most useful types of laser, known for its relatively high efficiency and high power. Solid-state lasers are a second type, with the most common being Nd:YAG lasers, also first demonstrated in 1964. These are more suited for fine details (they use a crystal made of neodymium-doped yttrium aluminum garnet, hence the abbreviation). In 1971, $CO_2$ lasers began to be used to cut wooden shapes; application to metals came in 1980, once it was discovered that metal cutting requires depolarized laser beams. Since then, CNC operation, increased power, and decreased cost have been the main drivers behind the strong growth of laser cutting applications. For the near future, this trend is set to continue, with laser cutting becoming even more common.

## Additional information: notes on cost

With such a wide range of shapes, sizes, materials, and series sizes available, there is no such thing as *the* cost of laser cutting. One rule, however, always applies: the thinner the material, the faster the laser will cut, and the lower the price per component will be (using a more powerful laser you could, of course, cut thicker materials at the same speed, but at the price of higher machine costs). Different materials also offer significantly different cutting speeds, with stainless steel being relatively easy to cut, followed by aluminum, then low-carbon steel, then titanium.

## Competing processes

Laser cutting may be steadily replacing mechanical cutting in industry, but this does not mean that it will soon have no competition. Aluminum, for instance, is so reflective (for infrared light) and conductive (for heat) that lasers are not effective on this common light metal (this problem obviously also comes up in laser welding of aluminum), whereas most plastics are too flammable. For these materials, abrasive water jet cutting often is the better choice. And for very heavy-duty cutting—such as for steel sheets that are 20 mm thick or more—plasma cutters will do the job at much lower costs, as the capital investment for the required ultra-powerful lasers is prohibitive.

## 13.5 ROTATIONAL MOLDING
### At-a-glance

This principle is ideal for shaping thermoplastics into large, thin-walled hollow shapes, provided the production volume is small and tolerances need not be very tight. Typical applications include canoes, buckets, waste containers and portable lavatories, but also several iconic design products, such as the Magis puppy–shaped children's chair (Figure 13.5). The principle operates as follows. A hollow mold defining the outer shape of the product is charged with a measured load of plastic pellets, then heated and rotated around two axes simultaneously, causing the pellets to melt and spread over the mold inner

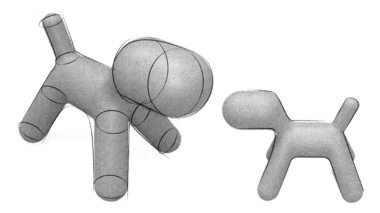

**FIGURE 13.5**

Magis puppy chair. (For color version of this figure, the reader is referred to the online version of this chapter.)

surface as a thin layer. Heating is usually done in a large hot-air oven. Once the mold emerges from the oven and cools down, usually with the aid of cold air or a water mist, the molten plastic solidifies and the product can be de-molded (plastics have much higher thermal expansion coefficients than common mold materials, so de-molding usually is easy). In practice, heating, cooling, and mold cleaning/charging are done with three molds in parallel to speed up the cycle time. This way, the "Magis puppy" takes just 25 minutes to produce.

Rotationally molded shapes are generally closed hollows, but subsequent cutting or machining can produce open, shell-shaped parts as well. Size falls within a range from a few decimeters to around 3 meters, with wall thicknesses of 2 to 8 mm. In theory, any thermoplastic can be used, but in practice, the majority of products are made of PE and ABS. With cycle times of 30 to 90 minutes for one mold, the process is slow, and hence its applicability to small production volumes, typically 100 to 10,000 units/ year. Control over tolerances is quite poor: predicting the wall thickness distribution and shrinkage over a product is difficult, so some warping and distortion is inevitable—and the small production runs usually leave little room for optimization through trial-and-error.

## Process history

The history of rotational molding actually predates man-made plastics. It was originally invented around 1850 by the Englishman R. Peters, who applied it to make metal objects. Several decades later, the process began to be used for hollow chocolate products—an application that is still in use today. Not before the 1940s was it finally applied to plastics, which had just become available: first PVC, and since 1958, PE. Today, the trend is toward more mechanization, increased complexity, and better understanding of the process fundamentals.

## Mold types and costs

Most molds for rotational molding consist of two halves and are of the simple "open-close" type, as the limited production volumes generally prohibit the use of slides or other complex mold features (the Magis puppy is among the exceptions, as it uses a four-part mold). The use of such molds implies that shapes should not have undercuts, but instead have a few degrees of positive draft angle all around. Furthermore, a parting line will be visible on the product, as will be the place where a pipe is placed to vent off the gases that are generated during heating. Extra effort is required to remove these surface defects. On the plus side, the product assumes the surface texture of the mold, which can range from smooth to coarsely structured. It is also possible to use inserts, such as metal connecting points, or to mold different layers over each other in sequential production cycles.

Molds can be made in a wide variety of ways and materials, depending on the size, shape, and quality requirements. Those for simple, large, single-curved shapes can be made cheaply with low carbon steel or stainless steel sheet material. Molds for more complex shapes require either sand casting or integral machining of aluminum, which is costly for large products but well suited for smaller ones; machining has the added bonus of close control over the tolerances. Because making molds is usually "one-off" production, it relies largely on manual labor, and we should expect mold prices to vary from one continent to the next. To give a rough idea of the investment level needed, in one feasibility study for a rotationally molded 75 cm tall PE bar stool, a leading Dutch production company quoted a price of 13,000 euro for the cast aluminum mold.

### Competing processes

Shapes that are rotationally molded can also be made by thermoforming (for closed shapes, two thermoformed halves can be joined) or by extrusion blow molding, at least in principle. Apart from the reduced thickness, there are, however, two subtle differences. In rotational molding, it is the corners that end up a bit thicker than the material in between them, whereas in thermoforming and blow molding, this is exactly reversed. This may give rotationally molded parts a certain extra ruggedness that comes in handy for everyday products. A second difference is that unlike thermoformed products, which started off as extruded sheet, rotationally molded products are essentially free from internal stresses, and therefore more stable in the long term.

## 13.6 EXTRUSION OF THERMOPLASTICS
### At-a-glance

In Chapter 5 we discussed extrusion of metals. The process also exists for thermoplastics, but it is so different physically that it should be seen as a separate principle. Underlying this difference are certain physical properties of the materials: the higher heat capacity and much lower heat conductivity of plastics as compared to metals, and the fact that molten thermoplastics exhibit visco-elastic behavior.

The bulk of plastic extrusion worldwide is aimed at semi-finished products, such as sheets, foils, and pipes. These are then shaped into finished products by other means: for instance, garbage bags are first extruded as a thin-walled hose, then blown into bags and heat-sealed on one end with the use of large machines with throughputs of over 10 ton/hour! Extruded pipes can also be the input for blow molding, which by itself is similar to thermoforming—the combination, known as "extrusion blow molding", is used for countless types of plastic bottles. Another common semi-finished product is electrical wire, where extrusion is used to coat the metal core with a plastic skin. However, the considerable form freedom of plastic extrusion also allows the manufacture of finished parts, such as PVC window frames,

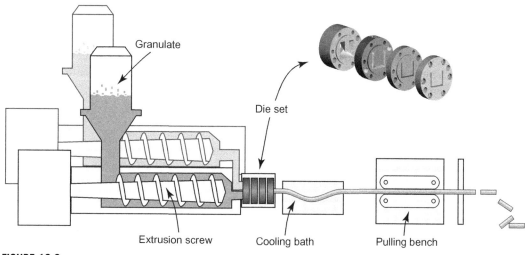

**FIGURE 13.6**

Schematic for co-extrusion, using a set of four die plates for a U-profile. (For color version of this figure, the reader is referred to the online version of this chapter.)

rubber window seams of cars, trim for furniture, cable guides, and so on. This "shape extrusion" (as opposed to bulk extrusion) is what is presented here.

A plastic extruder is similar to the injection unit of an injection molding machine, taking in plastic granulate on one end and ejecting a continuous stream of hot plastic on the other, typically 50 mm thick (Figure 13.6). As in injection molding, the pellets are heated through friction via a rotating screw, but now only to a temperature below the $T_s$ (for amorphous plastics) or $T_m$ (for semi-crystalline plastics), comparable therefore to thermoforming. The hot plastic is forced through a die consisting of a number of die plates, gradually forming the round stream into the desired profile. The number of die plates depends on the profile width, wall thickness, and type of plastic, and it can range from just two plates for a simple thin-walled tube to up to ten plates for wide, complex profiles.

When the profile emerges from the last die plate, it must be calibrated by sucking it firmly onto a cold calibration die using vacuum (except for relatively soft plastics and rubbers, which do not require this step), causing the surface to cool. Next, it is led through a water bath for additional cooling, stretched, and finally cut to length. Because the process is continuous, this last operation requires a movable saw. Compared to extrusion of aluminum, all of this is relatively complex, requiring more careful design and preparation and a more costly set of die plates for apparently the same profile. However, the process has certain unique strengths: for instance, extrusion-grades of plastics are tougher (i.e., better impact strength) than the accompanying injection molding grade. Also, it is possible to co-extrude two or more different plastics simultaneously.

## Process history

Plastic extrusion dates back to the 1930s and gained importance with the rise of the consumer society after World War II. In this, its history runs parallel to injection molding. Initially, extrusion was limited to single materials, but by 1960 industry began experimenting with co-extrusion. In this method, two separate injection units are combined to produce one product consisting of two components, or "2K" for short (as shown in Figure 13.6). These can simply be two colors of the same plastics, but combining different plastics is also possible, provided that their processing temperatures are compatible. Today, up to five components can be co-extruded. Current developments focus on more complex profiles with closer tolerances, more automation, additional processing steps in line with the extruder (e.g., labeling, painting, machining), and the application of bio-based plastics, such as PLA.

## Materials, capabilities, and limitations

Extrusion of thermoplastics is best done with amorphous materials, notably PVC, PS, LDPE, ABS, and thermoplastic rubbers, such as EPDM. Semi-crystalline plastics (e.g., PP, HDPE) can also be extruded, but their more pronounced shrinkage is a problem. In this respect, the process is comparable to thermo-forming. PC, although amorphous, poses special challenges and is only rarely extruded, whereas PA (nylon) is not extruded at all.

As for size, most suppliers can easily produce profiles with cross-sections measuring up to $20 \times 50$ cm, either solid or hollow. Extrusion speeds typically go up to 12 m/minute. Going larger or faster is not in itself a problem, because unlike extrusion of metals, press force is not the main limiting factor, and extrusion ratios do not apply either, as they do for metals. What *does* put a limit is the visco-elastic behavior of plastics. During the process, the plastic is compressed in the directions perpendicular to the extruded profile, and upon exiting the die, the profile will expand significantly—a phenomenon referred to as "die swell." This is routinely corrected by making the die smaller than the profile should be, but accurate predictions are difficult, especially for complex profiles. Die swell is also time and temperature-dependent, so faster cooling and, hence, faster production, will change the amount of distortion, with later relaxation of stresses. (This effect also plays a role in injection molding, but in extrusion it is much more pronounced.) Figure 13.7 shows how it affects simple shapes.

Another limitation is posed by the cooling speed. Too-rapid cooling of the profile surface will cause thermal stresses between surface and core, and, hence, distortion, surface defects, or even cracks. This is enough of a problem that a strict design rule is needed: the profile *must* have a constant wall thickness, which also helps with control of die swell. So for this process, function (profile shape, size, and material), cost (speed), and quality (tolerance, surface quality, stability) are closely linked.

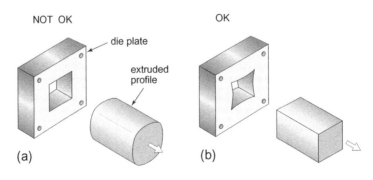

**FIGURE 13.7**

Die swell in extrusion of plastics. For a square profile, design (a) will not do; instead, design (b) should be used. (For color version of this figure, the reader is referred to the online version of this chapter.)

## Competing processes

For comparatively simple profiles, such as the rubber or plastic pipes, strips, or trim, extrusion really has no feasible alternative. Semi-structural profiles, of course, get competition from aluminum extrusion, whereas complex parts requiring a lot of machining after extrusion may be more economical if produced by injection molding. As always, the best choice depends on a careful balance between aspects of the manufacturing triangle—notably production volume against shape and material.

## 13.7 COMPRESSION MOLDING
### At-a-glance

Compression molding is a relatively simple process that can handle various types of materials but that is mainly used for shaping thermosetting polymers. A pre-measured amount of partially polymerized powder is placed in a hot mold, which is then closed for several minutes to allow the material to react, with temperatures of 200°C or more and pressures of several hundreds of bars. Alternatively, a soft, pre-formed shape can be used instead of powder (Figure 13.8). It delivers fully finished products, except for flash around the parting line that needs to be trimmed off. Compression molding can produce parts in a wide size range, from tiny kitchen utensils to car hoods, with production volumes anywhere between thousands and hundreds of thousands of units per year. Most heat-resistant plastic utensils in your house are compression molded, as are most of the plastic interior panels in trams and trains, to name just a few of the applications. Polyurethanes and polyesters are the main plastics used, but (thermosetting) rubbers are also common (e.g., rubber keypad components).

### Process history

Thermoset compression molding goes back almost as far as the materials themselves. It was the main method for shaping the first man-made plastic, Bakelite, in the 1920s. By 1960, the process was applied to glass-fire reinforced thermosets, leading to the use of *sheet molding compound* (SMC) and *bulk molding compound* (BMC, also known, somewhat confusingly, as "DMC"). These materials can replace, for instance, sheet metal in cars—notable examples include the 1984 Renault Espace and Pontiac Fiero cars. SMC technology and compression molding of unreinforced materials have continued to develop, improving its applicability, cost, and environmental impact.

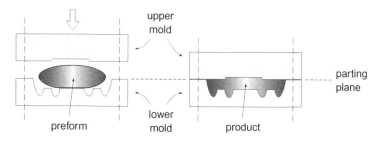

**FIGURE 13.8**

Process schematic for compression molding. (For color version of this figure, the reader is referred to the online version of this chapter.)

## Design detail and cost

In many ways, compression molding is comparable to injection molding of thermoplastics, but with some key differences. Parts cannot be as complex (undercuts, for instance, are even less suitable), but they can be larger, with at least equal control over tolerances and generally better stability and homogeneity, thanks to the absence of anisotropic flow patterns, knit lines, and other flow defects. Molds are less costly and molding machines are simpler, but as cycle times are longer than in injection molding, the total machine costs are comparable if not slightly higher. Part thickness affects cycle times, but not as strongly. The cure speed depends much more strongly on the material type and the processing temperature.

## Competing processes

For solid products made from thermosetting plastics, compression molding can be considered to be the main manufacturing principle, with little competition. For smaller production volumes involving fiber-reinforced plastics, RTM becomes competitive; for larger production volumes and more complex shapes, injection molding will often be the better choice.

## 13.8 PRESS-BLOW MOLDING OF GLASS

### At-a-glance

Glass is a mixture of oxides, based on silica $SiO_2$, with different types depending on composition. For example "soda-lime glass" is silica with around 15% NaO ("soda"), 8% CaO ("lime"), and a few % $Al_2O_3$ (alumina). This is the most common type, used for windows, jars, bottles, light bulbs, and so on. Replacing the lime with $B_2O_3$ (borax) yields a glass that is more difficult to process but more resistant to thermal shock. Sold under the trade name Pyrex, this "borosilicate glass" is used for chemical equipment, oven dishes, and quality coffee pots. A third type is "lead glass", with some 40% silica and up to 50% PbO (lead oxide), used for crystal glassware. Like all glasses it is fully amorphous and owes the term "crystal" solely to the luster of such products. Other types include glasses for reinforcing or optical fibers. Apart from being transparent and chemically resistant, glasses are hard and strong, with some glass fibers attaining strengths up to 4,000 MPa. Their stiffness too is good: the Young's modulus of soda-lime glass, for instance, is comparable to that of aluminum. Their main mechanical limitation is that they are very brittle. How can they be shaped into parts and products?

Melting and then blow forming is the long-established manufacturing method for hollow glassware. Modern glassware is now made slightly differently, through a process called *press-blow molding*. A piece of hot, soft glass (the "slug") is first pressed into a preform shape over a mandrel inside a cooled steel mold, then it is transferred to a second mold in which it is blown into the final shape using air pressure (Figure 13.9). After the product is ejected, trimming unwanted material is more difficult than for polymers and metals. The method's main benefit is that it can produce a smaller wall thickness than ordinary blow forming (with or without molds); also, it allows for modest thickness variations and details such as screw threads. Essentially, the process is similar to that used to produce the ubiquitous PET bottles: first preforming a parison by injection molding, then finishing through blow molding. Press-blow molding of glass is used for drinking glasses, jars, bottles (and all other kinds of glass

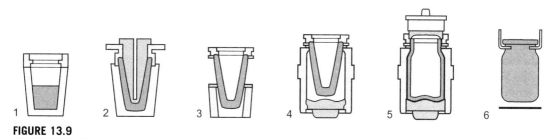

**FIGURE 13.9**

Steps in press-blow molding of glass: (1) deposit of the gob, (2) pressing of the gob into a preliminary shape, (3) and (4) transfer of the preliminary shape to the blow mold, (5) blowing the product into its final shape, (6) removal of the product to a conveyor. (For color version of this figure, the reader is referred to the online version of this chapter.)

containers), glass lampshades, light bulbs, and so on. When fully automated, it can reach staggering production volumes in continuous production: for instance, the Dutch Royal Leerdam glass factory produces 150 *million* drinking glasses per year. (Interestingly, this factory rotates the glasses during cooling, eliminating unsightly parting lines.) For small production volumes, the method it is not well suited, other than by specialty suppliers.

## Process history

Glass has a high melting point of around 1,300°C, but this temperature was already attainable with the charcoal-fired ovens of antiquity. So-called free-form blowing, in which glass blowers shape the material using their powerful breath and skill, requiring no mold, dates back to ancient Egypt. Glass processing using molds can be traced back at least as far as the Roman era, when wet molds made of pear wood were used. During the Industrial Revolution, press-blow molding was mechanized and scaled up. Initially, the actual blowing was done by humans, but since the 1850s in industrialized countries, this step has been automated to reduce costs (in low-wage countries today, even large production volumes of glassware are still blown by humans). Developments include scratch- and impact-resistant forms of glass, triggered largely by the increasing use of glass in portable electronics: the well-known "Gorilla glass" from Owens-Corning actually dates back to the 1960s, but returned to production in 2006.

## Design and cost

In spite of the high processing temperatures, glass is surprisingly easy to shape for such a brittle material. Because it does not crystallize, the problem of solidification shrinkage, often so problematic in casting of metals, is absent. This allows glassware to have some variation in wall thickness, such as jars or bottles with thickened corners. In press-blow molding, this can be achieved by smart forming of the slug in the first process step and careful placement of the air inlet in the second. Alternatively, keeping the wall thickness uniform is also possible. This is facilitated by the fact that during blow forming, thinner sections cool faster than thicker ones, getting more viscous and, hence, shifting further deformation to the thicker sections that are still hot. The material's low coefficient of thermal expansion (just 9 μm/(m.K) at RT—less than carbon steel) is another processing bonus. Still, glass is very brittle, and glassware should not have abrupt changes in wall thickness, nor sharp corners, to prevent breakage during handling.

For large production volumes, the cost of press-blow molding of glassware is largely determined by the base material price, also because the lifetime of the steel molds is comparatively long. For ordinary soda-lime glass, typically containing around 50% recycled material (not production scrap, but post-consumer scrap), this price is just 1 to 1.2 euro/kg, which explains why bottles, jars, and so on can be so cheap. Glass can be either colorless or colored: the latter only requires the addition of small amounts of other oxides (e.g., $FeO$ for green, and $MnO$ for blue) that need not significantly affect cost (though they complicate recycling: getting the pigments back out again to make clear recycled glass can require heavy use of chemicals). At 3 to 4.5 euro/kg, Pyrex is more expensive.

## Competing processes

Press-blow molding of glass has quite unique capabilities and hence, receives little to no competition, though of course there is intense competition for market share of the products themselves, such as beverage containers. Simple hollow re-entrant shapes, such as milk bottles, can alternatively be made economically using straightforward mold blowing, but this is rare, as the edge then still needs finishing. For solid shapes without undercuts that can be molded with a two-piece mold, such as plates or bowls, compression press-molding is more straightforward.

## 13.9  SLIP CASTING OF POROUS CERAMICS
### At-a-glance

Ceramic dinner plates, and similar products that are non-re-entrant, shallow, and solid can be press-formed from clay, then fired in an oven. Vases and other containers, being rotationally symmetric and hollow, can be turned on a potter's wheel. But how can we make large ceramic products that are hollow, deep, and not axially symmetric, such as a sink or toilet bowl? The answer is, by slip casting. In this common process, a suspension of clay particles—the *slip*—is poured into a mold made of plaster of Paris (i.e., gypsum). The gypsum immediately begins to absorb water, causing a layer of clay to form on the inside of the mold. Once this layer is sufficiently thick (typically after 20 minutes for a 5-mm thickness), the mold is tipped upside-down and suspended for a few minutes, draining off the remaining slip. It is then tipped back up and the shell is left to dry inside the gypsum mold for an hour or more, until it has attained a leathery consistency. Then, this "greenware product" is de-molded, smoothed, and often glazed before it is fired in an oven, where it obtains its final strength, hardness, and surface quality. During firing, the product shrinks significantly (up to 20%), with consequences for design, such as requiring generous radii and a constant wall thickness. After drying, the mold can be reused for a new production cycle. Slip casting is typically used for products weighing between 0.1 and 20 kg with wall thicknesses of 1 to 20 mm. In terms of production volume it is very flexible, ranging from small series (or even one-off products) to thousands of units per year in large factories (where, given the long drying times, many molds are used in parallel). Nowadays, cycle times can be reduced considerably by the use of porous polymer molds and by applying up to 40 bar of pressure on the slip. Dimensional accuracy is limited: for a typical slip cast bathroom sink that is nominally 50 cm across, the dimensions vary by several mm.

## Process history

Slip casting has been in use since antiquity. Around 200 BCE, the 8,000 soldiers of the Terracotta Army (Shaanxi province, China) were manufactured in this way, with the limbs of the figures cast separately prior to assembly. We can safely assume that in those days, many common utensils were also slip cast, and in considerable volumes; this certainly was true by the Middle Ages. Today, the scale of operations has shifted from small, regional factories employing many people to huge, almost fully automated operations producing millions of products per year. Small artisanal factories have only survived by switching to small-volume, high-value specialty "heritage" products, such as the world-famous Delft blue pottery, still made in the traditional way at the *Porceleyne Fles*.

## Competing processes

For non-rotationally symmetrical hollow shapes, slip casting of ceramics receives no competition, as it is practically the only way for making such products in one piece.

## 13.10 SURFACE HEAT TREATMENT OF METALS

### At-a-glance

Previous chapters have already presented three important heat treatments for metal components: Chapter 4 mentioned annealing of sheet metal (softening the material and restoring ductility, enabling additional forming steps), Chapter 5 discussed precipitation-hardening of aluminum (increasing strength), and Chapter 6 covered quench-temper hardening of carbon and alloy steels (again, increasing strength and hardness). These are the main *bulk* heat treatments, but designers using steels also need to know about *surface heat treatments*, sometimes known as "case hardening." Surface heat treatment allows us to give a steel component (such as axles, gears, ball bearings and ball races, or injection molding dies) a hard, wear-resistant surface, while retaining toughness in the bulk of the material. Fatigue strength can also be increased, as case hardening can leave compressive stresses in the surface that delay fatigue crack initiation. Again, many different types of surface heat treatments are available that exploit quite different metallurgical principles at work. Three common treatments are presented next to give you a feel for the subject.

### Carburizing

This common process operates by diffusing more carbon into the surface of a low carbon (typically 0.2% to 0.3%) steel component, to a depth in the range 0.2 to 1.5 mm, depending on processing time and temperature (which must be in the austenite region, typically around 910°C). By quenching the component, the surface turns to hard martensite (with a hardness equivalent to a yield stress of 2,000 to 4,000 MPa). Only the surface forms this brittle phase, due to the faster surface cooling rate and the increased hardenability due to the increased carbon content—the underlying bulk microstructure (and properties) are unaffected. Alternatively, the surface of the component can subsequently be "transformation-hardened" by a local surface heating and quenching cycle, again producing a thin surface layer of hard martensite with enhanced carbon content. (This is also used, without prior surface carburizing, for case hardening higher carbon steels and cast irons.) Another process variant is carbo-nitriding, which uses diffusion of carbon and nitrogen at lower temperatures of around 780°C, giving less distortion but also a thinner high-hardness layer (0.07 to 0.5 mm).

## Nitriding

A second common surface heat treatment, nitriding operates through the diffusion of nitrogen into the surface of a metal component, usually steel but sometimes also titanium or aluminum. It has an even lower working temperature than carbonitriding (around 550°C) and requires no subsequent quench-temper steps, making it applicable to more complex shapes. Nitriding can give the same hardness as carburizing in a 0.1-0.7 mm thick layer and is particularly used for steels containing alloy elements that will react to form hard nitrides (such as aluminum, titanium, and vanadium)—that is, it is used with the alloy steels selected for complex automotive parts and for the dies and molds used in injection molding and extrusion dies. As the process temperature is also suitable for tempering, the quench-temper bulk heat treatment can sometimes be conducted simultaneously with surface nitriding, thereby reducing costs.

## Induction hardening

In induction hardening, a copper coil carrying a strong, alternating current is placed around the part to be hardened (or traversed over it, if a large part requires treatment; Figure 13.10). Magnetic induction produces eddy current heating concentrated in the part surface, which literally glows red-hot in a matter of seconds. When the current is switched off, the component is quenched with a water spray or dropped into a waterbath; or if a traversing coil is used, a ring-shaped water spray tracks along behind the coil.

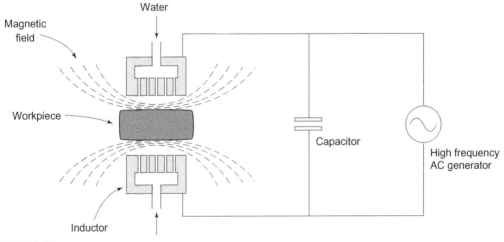

**FIGURE 13.10**

Schematic of induction hardening. (For color version of this figure, the reader is referred to the online version of this chapter.)

With rapid surface heating and a quench, induction hardening is used to harden only the surface to martensite, typically for steels around 0.4% carbon, giving a Vickers hardness between 400 to 700 HV in a 0.5–5 mm thick layer). The same heating method is also used for the complete quench-temper bulk treatment, by leaving time for heat conduction through the thickness to fully austenitize the part, which is more efficient than furnace heating for small parts. A benefit of induction hardening over the previous diffusion-based methods is that it is possible to treat just parts of the component's surface rather than the whole (an advantage that also holds for other transformation hardening techniques, such as laser hardening of steels).

## 13.11 COATING, PAINTING, AND PRINTING PROCESSES

To conclude our discussion of manufacturing processes, we look at the broad range of surface processes that are used to finish a product: coating, painting, and printing. As usual, we only give a brief summary of the subject by discussing processes that are especially important or interesting and pointing to the design factors to think about for all finishing processes. We can, however, make one generic remark: although they are based on different physical or chemical principles, these processes all have one thing in common—something is *added* to the surface, usually from tens to hundreds of micrometers thick, that changes the part dimensions. In this, the processes differ from the surface heat treatments discussed earlier, which modify the microstructure below the surface. Surface finishing processes are applied to metals, plastics, and ceramics, and they change the functionality or look and feel of the product (or both)—that is, its resistance to corrosion, scratching, or wear, and the presence of color, texture, or patterns. Applying the right coating, paint layer, or print pattern can determine a product's success or failure, so every designer should become familiar with at least a handful of processes.

### Coating processes

Under coating processes we find anodizing, electroplating, hot-dip coating, and many more. Here are some brief notes on these three.

*Anodizing* is commonly applied on wrought aluminum (e.g., extrusions) to increase its surface hardness, and corrosion resistance, and also color: if so desired, the process can deliver a surprising array of colors and a smooth, high-tech finish (popular for architectural applications). On aluminum castings it is less effective, giving a hard and corrosion-resistant but somewhat mottled look.

*Electroplating* is a flexible process that is applicable not only to metals (e.g., silver-plated cutlery) but also to ceramics or plastics, provided they are coated first with a thin, electrically conductive layer. The process operates through electrolysis with ion exchange at the surface of an immersed product. This means that surfaces inside of hollow parts can be left uncoated (electric charges repel each other toward the outer surface of such parts). Common chrome-plated parts, from automotive trim parts to steel bicycle components, are coated with this process.

*Hot-dip coating* is applied specifically to steel parts and products, which are dipped in a bath of molten zinc to make them corrosion resistant. This is known also as "galvanizing" and gives a tell-tale appearance in which large individual zinc grains are clearly visible, as on five-bar gates.

## Painting processes

Again, a single term is used to cover a range of individual processes. Important painting processes to know about are electropainting (or electrophoretic deposition), solvent-based painting, and water-based painting. What they have in common is that paint is spread over the surface of the product and left to dry, so they all use solvents that evaporate and allow the paint to solidify, usually through a thermosetting reaction. A fourth, very different painting process, is enameling. Again, here are a few words on each.

*Electropainting* is the process of choice for the under-coatings of car bodies. These are immersed in a bath and electrically charged, causing the oppositely charged paint particles to deposit onto the surface. Next, the part is taken out of the bath and left to dry, usually at elevated temperatures (car bodies are dried for around 30 minutes at 200°C). A single electropainting step can meet functional and aesthetic needs for many products, but for cars, it is followed by one of the next two processes.

*Solvent-based painting* uses a suspension of pigments and resins in a volatile organic solvent (VOC), which is brushed, dipped, or sprayed onto the part, then the part is left to dry while the solvent evaporates. It is highly developed and flexible, producing paint layers that are functional and aesthetic in a single step (e.g., "soft-touch" paints). The environmental concerns inherent to the use of VOCs are causing a shift toward the next variant.

*Water-based painting* is similar to the previous method, with water used as the solvent. Surface quality is usually not as good and drying times are longer, but the paint industry is hard at work to fix these problems.

*Enameling*, simply put, is "painting with glass." Very high temperatures are necessary to melt the glass, which restricts the use of this method to certain metals and ceramics, but the results are spectacular, including exceptional hardness and corrosion resistance and a full range of lustrous colors.

## Printing processes

The difference between painting and printing is simple: with printing, instead of the whole product, only parts of it receive a coating. This allows you to put logos onto a product, or instructions, numbers, symbols, and so on. There are several dedicated methods for doing this, such as cubic printing, pad printing, and silkscreen printing.

## 13.12 STILL NONE OF THE ABOVE?

We have now covered a wide range of principles and methods for all material classes. There are, of course, many more: superplastic forming of metals, metal injection molding, dipping of rubbers, steam bending of wood, autoclaving of fiber composites, and so on. But you should now be equipped to approach *any* process with your eyes open, armed with the manufacturing triangle to guide your thinking. A sensible starting point with an unfamiliar process is always a bit of lateral thinking—To which principle is this process most alike (in terms of material, form, application area, etc.)? Then remind yourselves of which issues in the triangle were dominant in this case. There are many sources of information at your disposal when you want to explore new processes—or to enhance your insight into any of the processes covered in this book. This final section gives you some guidance on sources.

## Books

University textbooks on manufacturing tend to be encyclopedias containing nearly all conceivable processes. No reader can be expected to know everything that is in it, but this span of coverage does make them good references. On the whole, however, they often only describe the processes, instead of explaining them, leaving you with many facts but little physical insight. Another drawback is that some are written from the perspective of one class of materials—often metals, sometimes plastics—and are not as strong in their coverage of processes for shaping other materials. Textbooks can also be outdated: as a general rule, editions more than 10 years old should be used with caution. Two textbooks that do not have these issues and that we recommend are *Manufacturing, Engineering & Technology* (Kalpakjian and Schmid, 2005, Prentice Hall, 5th edition) and *Manufacturing with Materials* (Edwards and Endean, 1990, The Open University). The latter includes a set of reference cards for many processes and presents and rates several aspects of the manufacturing triangle in a systematic format.

Apart from textbooks, there are also works for professionals (many listed in the "Further Reading" section at the end of previous chapters). Again, check the age of the edition, though the value of information varies widely depending on how long-established the technology has become—casting handbooks from the 1970s contain valuable information, whereas a work on additive manufacturing from the year 2000 is more or less obsolete. Also, be wary of books written from a strong company perspective, where there is a hidden agenda to sell you something (i.e., the company's own processes). Furthermore, be prepared for specialist jargon and unfamiliar definitions (which may not always be as logical as one might hope).

Several books aim to introduce designers to the world of manufacturing by means of examples, with a focus on visuals and results instead of text and analysis. Good examples are *Manufacturing Processes for Design Professionals* (Thompson, 2007, Thames and Hudson) and *Making It* (Lefteri, 2012, Laurence King Publishing, 2nd edition), and we recommend using these books as a complement to our own.

## Computer databases

Several databases are available that offer a good overview of the world of manufacturing processes. We recommend the *Cambridge Engineering Selector* for its completeness and accessible user interface (which it shares with extensive material property databases). Database sources are ideal for browsing, selecting, and getting information, but they are not well-suited for building up insight: that requires different sources and a different approach.

## TV programs

One good way to see processes in action is via TV series such as *How It's Made* and *Megafactories*, which reveal the hidden world of manufacturing to the public. Such programs tend to focus on a specific product rather than process and are about entertaining and informing a lay audience. And, of course, all featured companies have a strong interest in appearing at their best. For instance, you may be shown all kinds of quality control operations in action, but learn little about how many products fail to meet specifications—let alone why, and what measure to take to improve the situation. In the end, TV programs can only give you an introductory, superficial feel for what large-scale manufacturing operations are about.

## The Internet

Most of the comments on TV programs also apply to what you can find on the Internet. YouTube has become indispensable for finding video material of processes in action—not all of equal quality! And the ubiquitous Wikipedia does contain a wealth of information, but the material is anything but complete: well-established processes can get surprisingly little coverage compared to more recent developments (in this sense, the Internet is almost the reverse of books). Moreover, most of what you can find is not vetted in any way and may be incorrect, or it may be posted by people with an agenda to sell their process or material—in other words, the information is not necessarily provided by objective, independent experts. So you need to be wary of what you read online (in Wikipedia, the "talk" page can be worth checking). The Internet is vulnerable to insufficient maintenance, with many good initiatives becoming defunct in only a few years if they are not maintained. Good examples, which it is hoped will have longevity online, are www.productionnavigator.com, www.manufacturing.stanford.edu/hetm, and www.bencollette.com/productsbyprocess.

## Scientific publications

If you have access to scientific databases (as any major university library can grant you), one source of information you should not rule out is the academic, scientific literature. These sources tend to be *very* specific and scientifically advanced, but if you have sufficiently specific objectives in your research, they can give cutting-edge information and insight you are unlikely to find through different channels. Some pointers: look for authors working in research centers that specialize in the relevant materials or processes, with industrial coauthors, and for work that describes real experiments or industrial data (not just simulations).

## Professional publications

Less formal, but more accessible, are the various journals and publications made by professional design or engineering societies, typically available at a national level. Examples are the German magazine *Kunststoffe* (German for "plastics"; see www.kunststoffe-international.com) and *Modern Casting*, issued by the American Foundry Society (AFS, via www.afsinc.org). The more influential organizations behind these publications often organize conferences in their fields, including materials, manufacture, and design.

## Trade fairs

Most industry branch organizations, such as the AFS (mentioned earlier), hold regular trade fairs. These events are ideal for giving attendees a feel for the latest developments and applications, often showing processes in action and, best of all, allowing face-to-face meetings with specialists. Examples are the *Aluminum Messe* (aluminum processes and applications), held in Essen, Germany, every three years, and the famous *"K"* (from "Kunststoffe"), held in Dusseldorf, Germany, every four years (this second fair is *huge*). Smaller, national-scale events are also well worth a visit. These are essentially commercial events, so, of course, each company will only present its most favorable face.

## Company visits

We have saved the best source for last: visits to actual manufacturing companies. This is hands down the best way to get acquainted with all the levels of manufacturing, down to the equipment itself. Company visits allow you to see the actual production at its actual pace, warts and all (unlike the demonstrations and videos you see at trade fairs or on television). Ideally, it is best to go with an actual part or product in mind that the company could manufacture for you, as this gives a focus for discussing process capabilities, such as available sizes, mold costs, tolerances, and so on, but also the dependencies among properties (i.e., the trade-offs between aspects of the manufacturing triangle). Again, be aware of the difference between "that cannot be done" and "that cannot be done by us", as the company is unlikely to refer you to a competitor! It is important to be able to distinguish what is (im)possible in principle from what is (im)possible in practice, and between what one supplier can do and what another one can contribute to your product design. Insight and understanding matter.

## Structuring your search: the manufacturing triangle

With such a wide range of sources available, you clearly need a way to structure the information you will find, and indeed, you'll need to know when to stop searching: When do you have "enough"? In this effort, you can make good use of the manufacturing triangle with its 15 attributes, grouped under function ("what can it do?"), quality ("how well can it do it?"), and cost ("what will be the price?"). Use these attributes as a kind of checklist and as a means to ask the right questions. For instance, if you encounter a new process, just begin by listing the different materials it can handle and the shapes and sizes it can produce. Next, look into the accuracy it provides, or the typical investment level it requires, and so on, working your way around the entire triangle. If you have covered all 15 attributes, your data set should be complete. But the analysis should not stop there: next it is time to look into the inter-dependencies between attributes. Look beyond what is obvious. For instance, larger parts will generally require higher investments than smaller ones and will be made with wider tolerances—at a qualitative level, this is nearly universally valid and therefore not very informative. The challenge is to get to the numbers and, from there, to dig into the underlying material science, into the "why." Taken together, the previous ten chapters provide you with a firm basis for doing so, with an extensive set of examples of what to look for.

During your search, you should always take into account that in manufacturing, science can only go so far in providing answers: sometimes things are simply done in a certain way because it made sense to do so at some point in the past, and they have since been "locked in" as an industry standard that may appear arbitrary, yet would now be very costly to deviate from. For instance, material science does not forbid having round instead of rectangular molds for injection molding, yet the use of the latter now dictates how things are done, and round molds would at the very least require a long lead time. Remember also that different suppliers will have different habits, preferences, and possibilities, and what is normal for one may be impossible for another. The world of manufacturing is anything if not practical, and although scientific insights can greatly help to show what can be done and why, they will never provide the full picture—but for that matter, neither would an approach based only on practice: you need both.

## 13.13 **Further Reading**

Taken together, the "Further Reading" sections of the previous ten chapters already give many hints on books you can use to continue your search (for instance, the book by Osswald and Menges mentioned in Chapter 8 may also serve as a reference for rotational molding and extrusion of plastics). Other books were mentioned earlier. We can add the following specific works for processes in this chapter:

Czerwinski, F., 2008. Magnesium Injection Moulding. Springer Science+Business Media, New York, United States. (*An excellent book on thixoforming. See also the website www.thixomat.com.*)

Lefteri, 2002. Glass, second ed. Rotovision, Hove, UK. (*Well worth reading, one of an excellent series of books on Materials for Inspirational Design.*)

Powell, J., 2008. $CO_2$ Laser Cutting. Springer-Verlag, London, UK.

# Recycling

# 14

## 14.1 INTRODUCTION

In essence, recycling is the reverse of manufacturing: products go in, and materials come out. It is well-established and highly developed, but is often misunderstood. For instance, the scale of recycling operations is routinely under-estimated, and although recycling usually has a beneficial effect on the environment, it is primarily an economic activity, done to make money. More importantly, although products are manufactured using specific production facilities, they are generally recycled in bulk, with different types of products ending up in the same waste stream and getting processed together. This chapter explains the basic concepts behind this huge industry (in the European Union in 2008 recycling amounted to 60 billion euros of material value) and aims to give you the understanding needed to design products that can be recycled efficiently. A key is knowing in which waste stream your product is likely to end up. Again, the text presents only an overview, but even this short chapter will equip you to be a critical reader of the many design-for-recycling guidelines in books or on the Internet.

The manufacturing triangle of function, cost, and quality can also be used to interpret recycling issues with products. Certain products can be easily recycled into separate high-quality materials, whereas others cannot (or only at high costs). The concept that explains this trade-off is the *grade-recovery curve*, which you will encounter in this chapter, together with other important terminology. Economic and ecological considerations are addressed next, and the chapter ends with discussions of the various waste streams that are recycled and reprocessed today. Our focus will be on recycling of post-consumer products. This is distinct from recycling of production scrap, and the two should not be confused.

## 14.2 RECYCLING TERMINOLOGY, STEPS, AND TOOLS

Figure 14.1 depicts the life cycle of materials (simplified, but sufficiently detailed for our purposes). We see that along with re-use and remanufacture, recycling is one of three different ways for materials to get a second life. Zooming in on recycling itself, notice how it consists of (1) collection, (2) liberation, (3) concentration, and (4) reprocessing. These four steps are explained here. If the recovered materials have a comparable quality to the original, "virgin" materials used in the product, we may speak of *recycling*; if quality is reduced (which is common, especially but not exclusively for plastics), we should use the term *downcycling*. Sometimes, all that can be done with the recovered materials is to burn them as

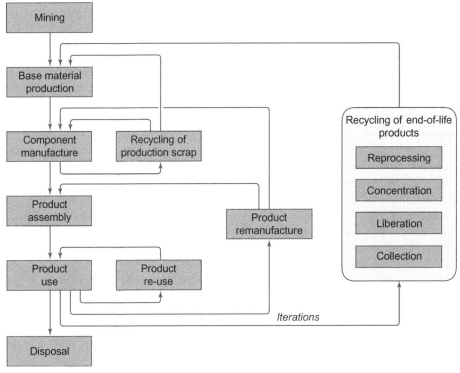

**FIGURE 14.1**

The life cycle of materials. (For color version of this figure, the reader is referred to the online version of this chapter.)

fuel (known somewhat euphemistically as *thermal recycling*). And to be complete, we can add that some plastics can also be reduced to their constituent monomers during recycling, which is then referred to as *back-to-monomer* (or *back-to-feedstock*) *recycling*.

One key aspect not captured in Figure 14.1 concerns time. Depending on the product, the period between manufacture and recycling ranges from weeks or months (e.g., disposable cups and packaging) to years (e.g., 5 years for most consumer electronics, 10 years on average for passenger cars). In expanding markets, this means that even full recovery of the materials by recycling can only partially meet the need for new materials in manufacture.

---

**EXERCISE 14.2**

Assume we have an average 1% (EU) versus 9% (China) annual market growth, and assume full recovery during recycling (impossible, as we shall see later). How much of the demand for new materials can be met by recycling, for product life spans of 5, 10, and 20 years, assuming that the amounts of materials used for individual products remain the same?

---

Another important aspect not shown in Figure 14.1 concerns losses in the life cycle of materials. Perhaps the largest of these is due to imperfect collection: not all products that could be recycled actually are collected, or if they are, they end up in the wrong waste stream. For an example, think of old cell phones and their chargers. These products spend a lot of time in a drawer somewhere while their owner uses a newer set, increasing the delay between use and recycling. But currently, just 5% of discarded phones go to the proper collection points—the rest end up in the domestic waste stream. As Section 14.6 will show, it is *much* easier to recover the copper and other valuable metals in these products if they are re-processed separately. Admittedly cell phones do have one of the lowest return rates, but with very few exceptions it is never 100%.

---

**EXERCISE 14.3**

Think of design strategies that facilitate the (separate) collection of end-of-life products, such as labels or electronic markers. Then consider the viability of laws to achieve the same objective. What is your conclusion regarding design for recycling?

---

Once products have been collected, usually multiple types together (e.g., automotive and white goods), step 2 is *liberation*. Prior to shredding, it is usual to partially disassemble the product to secure key components and materials. For passenger cars, this could include starter motors, tires, engine oil, air bags, and so on—essentially all items that are sufficiently valuable for re-use, or that can be reprocessed directly, or that represent a contamination hazard during subsequent operations. For most types of electronic products, partial disassembly is rarely done; for household waste, it is non-existent. It always involves manual labor, which sets the ceiling on how much time and effort can be expended. The bulk of the collected products are then shredded in large, crude "hammer mills" or similar machines, reducing it to chunks. Ideally, each chunk consists of a single material (e.g., steel, aluminum, glass, or rubber), but in practice there will be many mixed chunks of two or more materials, given the size range between individual parts and whole product, in a car for example. The chunks can be sorted by size using a set of sieves, with the larger chunks going round for a second time.

Next, in step three, the chunks are sorted into material fractions (steels, plastics, etc.), seeking to raise the concentration of dominant material in each pile of chunks. Sorting can be done by hand, by using machines, or by using a combination of the two. Obviously, hand sorting becomes progressively less economical as the cost of labor increases, but even in high-wage countries you still encounter it, particularly when a few misplaced chunks of a harmful material downgrade a large flow of valuable material. Examples are the removal of electrical motors (i.e., copper) from steel scrap before melting, and the removal of ceramic containers from waste glass prior to crushing.

---

### EXERCISE 14.4

The smaller the chunks, the greater the chance that they contain only a single material (i.e., they are fully liberated); conversely, larger chunks tend to be a mix of materials (i.e., they are not fully liberated). Why not shred everything into small chunks directly?

---

Provided they are well liberated, mixed metals can be sorted at low cost on the basis of certain material properties—this step is referred to as *separation*. For instance, magnetic separators use ferromagnetism, eddy-current separators use conductivity, and sink-float separators use density. Figure 14.2 shows these three common techniques in sequence to separate mixed metal streams. After concentration, most metals can be reprocessed and recycled without a significant loss of quality, assuming the degree of concentration, or "grade," is sufficiently high.

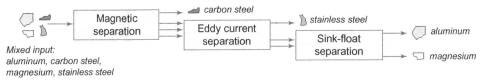

**FIGURE 14.2**

Three different types of separators in sequence, used to separate mixed metals. (For color version of this figure, the reader is referred to the online version of this chapter.)

---

### EXERCISE 14.5

Search the Internet for illustrations of the three types of separators shown in Figure 14.2. What are their typical dimensions? Hint: also use "sorting" as a search key word (e.g., "eddy-current sorting").

---

### EXERCISE 14.6

Why can magnetic separation be used to separate austenitic stainless steels from plain carbon steels?

---

Mixed plastics are not as easy to concentrate into separate plastic types (PP, PA, etc.): they are neither magnetic nor electrically conductive, and their similar densities rule out cheap and easy sink-float separation. But sequences such as that shown in Figure 14.2 can be extended with a higher investment level—for example, finely tuned sink-float separation can extract just the polypropylene from mixed polymers and other low density waste emerging from shredded cars. Setting aside several less common separation methods, this only leaves sorting of objects individually, by hand. To facilitate this process, it has become customary to designate the type of plastic on the part itself—indeed, in many countries, such labeling is now mandatory. Commodity plastics (LPDE, PET, etc.) have their own internationally accepted logos and codes in the form of stylized Möbius loops around a number code.

---

**EXERCISE 14.7**

Which material properties could, in principle, be used to automatically sort plastics?

---

Of course, bearing a designation or logo does not guarantee that the plastic part will actually be recycled. It merely facilitates hand sorting. But within a single designation or logo we would still find a huge variety of grades with different colors, fillers, molecular weights, and so on, and mixing these up inevitably leads to downcycling, not recycling. This is why downcycling is nearly always inevitable for plastics. Exceptions generally involve huge streams of identical products that are easier to liberate and concentrate, such as PET bottles—though even this material does not go back into bottles but goes into fibers for things like fleece jackets. And note that the claim "contains recycled material", seen on an increasing number of plastic products, rarely refers to post-consumer scrap but only to production scrap. This has already been recycled for years, so these claims are little more than a public relations exercise.

---

**EXERCISE 14.8**

From Chapters 3 to 11, select three different manufacturing principles. For each one, describe what the production scrap typically looks like.

---

Following concentration, the fractions are ready for the fourth and final step in recycling: reprocessing into a suitable semi-finished product. Metals can become new sheets, ingots, and so on; glass can be re-melted; and plastics can become new granulate. For some materials, this reprocessing is integrated with the production of virgin materials (i.e., making materials from ores), notably for steel and copper. For others, the virgin and recycled materials follow separate routes toward new applications. Here, aluminum presents an interesting example, with high-pressure die castings generally containing almost 100% recycled material, whereas wrought aluminum receives much less of this stream (largely due to the higher alloying level in castings, with lower sensitivity to trace impurities). Wrought aluminum producers are busy finding out what tolerance there is in processes such as extrusion to greater levels of impurities—the current alloy designations probably being over-specified but without anyone knowing by how much.

## 14.3 THE GRADE-RECOVERY CURVE

Now that we know the key terminology and process steps involved, we can take a closer look at recycling. Every recycling process can be modeled as in Figure 14.3. It has one input stream—the discarded products—and several output streams. If an output can be sold at a profit and reprocessed, it is called a "fraction"; if it can only be landfilled, it is called a "tailing" (both terms originate from the world of minerals processing).

**FIGURE 14.3**

Basic input-output model for recycling. (For color version of this figure, the reader is referred to the online version of this chapter.)

Crucially, the fractions are never 100% concentrated but always contain unwanted materials and contaminations. We can now define two key properties:

- *grade* (fraction X) = mass of material X in fraction X/total mass of fraction X
- *recovery* (material X) = mass of material X in fraction X/total mass of material X in input

*Grade* is a measure of quality and it captures concentration levels (i.e., how pure a certain fraction is). Theoretically, the market price of a fraction per ton goes up when its grade increases. In the pragmatic world of recycling, this relationship often takes the form of certain minimum values for the grade coupled to prices. The value then lies in what industry can use the fraction for, which is strongly affected by the influence of any contaminating elements. For instance, a steel fraction with less than 0.2% copper can be used directly for rolled steel sheet and is worth around 400 euros/ton (note that rolled steel itself costs 800 euros/ton). If the copper content is between 0.2% and 0.4%, then the fraction is too contaminated for sheet material but can still be used to make rebar, fetching around 100 euros/ton. Steel fractions with more than 0.4% copper basically have zero value because they must be mixed with pure iron to lower the copper content to an acceptable level. (Interestingly, steel with 20% or more copper actually *has* value again, as a source of copper!) Some other elements are at least as damaging as copper, whereas steel is more tolerant of several other elements (e.g., aluminum). Refer to the "Further Reading" section for more details.

If grade captures quality, then *recovery* is a measure of quantity: it describes how much of a certain material in the input stream is made available for reprocessing. A recovery of $R\%$ means that $(100 - R)\%$ of the material going into the process is lost, ending up either in the tailings or as a contaminant in one or more fractions.

Crucially, grade and recovery depend on one another. Specifically, if we want to increase the grade of a fraction of a certain material (for instance, to stay under a certain contamination level), we will reduce the recovery of that material, and vice versa. The dependency can be plotted as the *grade-recovery curve*, shown idealized in Figure 14.4. It exists for several reasons. One is that in practice, the various materials are never fully liberated. A second reason is that concentration is never perfect, especially if it must be done with cheap equipment, at high speeds, or both. So to improve grade and recovery simultaneously, we have to intensify our expenditure of money, energy, or both.

As a simple example, consider the recycling of PVC-coated copper wire. We have one input (the wire) and two outputs: the copper fraction and the tailings. After shredding, some bits of copper will

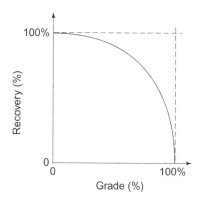

**FIGURE 14.4**

Idealized grade-recovery curve. (For color version of this figure, the reader is referred to the online version of this chapter.)

still have some PVC attached to them. Depending on the process settings of the concentration step, these bits may end up in the copper fraction (increasing recovery, but reducing grade) or in the tailing (reducing recovery, but increasing grade). With normal settings, the copper fraction consists of most of the copper plus a little PVC, whereas the tailings consist of most of the PVC plus a little copper. The tolerance to contamination is strongly influenced by what happens next—if copper containing a little PVC is remelted, the PVC will burn and be gone; the reverse is not true for the PVC containing some copper. For multi-material products such as cars, the situation gets more complex, as Section 14.6 will show. However, the principle remains the same, as does the trade-off between grade and recovery.

---

**EXERCISE 14.9**

By weight, copper wire contains some 70% copper and 30% PVC. Suppose we process 1 ton of wire per hour into a copper fraction weighing 0.74 ton/hour, of which 0.69 is copper and 0.05 is PVC. What is the weight and composition of the tailing? And what are the grade and recovery of the copper?

---

**EXERCISE 14.10**

How would extra-fine shredding of end-of-life products affect the grade and recovery of the product recycling process? What would be the extra costs?

---

The grade-recovery curve is usually applied to the two core steps in recycling (i.e., liberation and concentration). It can, however, also be extended to reprocessing, where there are certain unavoidable losses as well. For instance, when recovered metals are molten, some of the metal is lost through oxidation. This is known as "burn-off" and these losses increase with decreasing particle size (e.g., for aluminum, 1% to 2% is typical for larger particles, but around 7% is typical for production scrap from integral machining). Similarly, when recovered plastics are extruded into new granulate, they must be filtered to remove contaminants, with a small percentage of the plastic ending up in the filter and being lost to the cycle. So recycling always involves a compromise: the more we try to close the loop, the more expensive it becomes and the more energy it takes to do so.

---

**EXERCISE 14.11**

Imagine we design a car structure combining steel with aluminum (as several of today's cars already have). Assuming it will be shredded during recycling, what would be better: a few large fasteners or many small ones?

---

## 14.4 ECONOMIC ASPECTS OF RECYCLING

As stated in the introduction, recycling is primarily an economic activity, done for profit. If the value of the recovered materials exceeds the costs of recovery, then recycling makes economic sense. So let us now put some numbers to the four steps involved in recycling, see how things add up, and compare this result against the value of the recovered materials.

Step 1: Collection immediately presents us with a challenge. Household waste *must* be collected and recycled, as landfill options run out and rise in cost, but the total costs of collection and disposal/incineration are considerable: typically 300 euros/ton. A similar fee applies to industrial waste (provided it is not chemically active, which would make it much more expensive to collect safely). However, there is something of a market distortion in attributing collection costs to the recovered materials, as we have to incur these costs anyway for landfill disposal. On the other hand, certain products are collected virtually for free, because people voluntarily put them into separate bins (common for glass, beverage cans, and batteries). A third situation applies to cars and "e-waste" (household appliances, computers, etc.), for which there is a specific, growing recycling industry dedicated to material recovery instead of simply to waste management. For these products, the collection cost is significant and should be included in the overall recycling costs.

Whatever the situation, transport costs will be substantial: around 0.20 euro/(ton.km) for relatively dense materials and significantly more for low density material streams, such as waste plastics or discarded household products (which may fill up a truck's cargo volume long before its weight limit is reached).

Step 2: Liberation involves shredding, possibly preceded by partial product disassembly. The cost of disassembly obviously depends on the country in question, as it involves manual labor, and varies between 10 and 20 euros/hour for richer countries and just 1 to 2 euros/hour for some developing nations. The cost of shredding varies much less: typically between 25 euros/ton for coarse shredding (e.g., for cars) and 50 euros/ton for fine shredding (e.g., for plastic-intensive electronics, but sometimes also for steel scrap containing electrical motors, to liberate the copper wire from the steel parts). Note that the scale of liberation operations is huge: a car shredder can easily process around 100 tons/hour!

Step 3: Concentration, as we saw earlier, usually combines manual sorting and mechanized separation. Here, the cost involves labor costs and machine costs per hour, plus the amortization of all investments necessary to set up the process. Scale matters: as a rule-of-thumb, if the material stream to be concentrated is smaller than 1 ton/hour, then setting up a specific concentration facility is economically unattractive (exceptions involve either precious materials or dangerous substances).

With a typical two-shift operation, this equates to some 3,000 tons/year. Also, note that investment levels rise sharply with increasing demands on grade and recovery: in this sense, recycling suffers from a kind of reverse economy of scale. A case in point is the Dutch glass recycling history, for which this development is summarized in Table 14.1, which illustrates the costs of trying to close the loop!

**Table 14.1** Historical data on the investment level in glass recycling: the Netherlands

| Year | Separation Technologies | Investment (euros per ton/h) | Total Recovery |
|------|------------------------|------------------------------|----------------|
| 1972 | Washing, crushing | 5 | < 10% |
| 1982 | + Manual sorting | 16 | 10%-20% |
| 1987 | + Metal separation | 26 | 20%-50% |
| 1993 | + Stone separation | 40-65 | 50%-80% |
| 1998 | + Color separation | 80+ | > 80% |

Finally, step 4: Reprocessing involves costs that are too varied to capture and often cannot be separated from costs of virgin production (e.g., as for steel). Fortunately, these costs are usually not decisive, provided the grade of the material is sufficiently high.

Let us now look briefly at the value of the recovered materials. For five illustrative materials, the numbers are listed in Table 14.2 (along with the "gross energy requirements (GER)", which we will discuss in the next section). Of these materials, steel and copper are recycled without loss of quality and can therefore replace virgin materials one-to-one. The price of primary and secondary material is then similar. The other three are downcycled—wrought aluminum to castings, PP to low-grade applications, borosilicate glass to low-grade glass—and hence have lower secondary than primary prices.

**Table 14.2** Energy and price data for various primary and secondary materials

| Material | GER$_{primary}$ (MJ/kg) | GER$_{secondary}$ (MJ/kg) | Price$_{primary}$ (euros/kg) | Price$_{secondary}$ (euros/kg) |
|----------|-------------------------|---------------------------|------------------------------|-------------------------------|
| Low carbon steel | 29–35 | 8–10 | 0.45–0.55 | 0.40–0.50 |
| Wrought aluminum | 200–215 | 18–19 | 1.80–1.90 | 1.60–1.70 |
| Copper | 68–74 | 17–19 | 5.10–5.60 | 4.80–5.30 |
| Polypropylene (PP) | 85–105 | 36–44 | 1.40–1.50 | 0.80–1.00 |
| Borosilicate glass | 24–26 | 11–12 | 3.00–4.50 | 1.00–1.20 |

GER, gross energy requirements.

So how does this add up? To answer this question, we will consider a simple, common household appliance: a percolator-type coffee maker (Figure 14.5). It contains all five materials listed in Table 14.2. This Braun coffee maker contains 0.20 kg of steel, 0.10 kg of wrought aluminum, 0.12 kg of copper, 0.70 kg of PP, and 0.28 kg of glass, with a total weight of 1.40 kg. (It also contains small amounts of rubber and PVC, some ceramics in the heating element, and some solder; we ignore these here for brevity.)

**FIGURE 14.5**

Coffee maker. (For color version of this figure, the reader is referred to the online version of this chapter.)

*(Design: Braun).*

---

**EXERCISE 14.12**

What is the total potential recoverable value of the materials in one coffee maker in euros (use the data in Table 14.2, and give a range for your answer)?

---

**EXERCISE 14.13**

Estimate how many coffee makers are discarded in your country every year. Assume a 10 year product life span and one coffee maker (similar to the Braun) per four inhabitants. What is the size of this waste coffee maker stream in tons/year?

---

**EXERCISE 14.14**

Assume collection is done for free (e.g., people handing in the coffee makers voluntarily) except for transport costs. If we can use 10% of the material value to pay for transport, then what is the maximum transport distance we can afford?

---

**EXERCISE 14.15**

Full disassembly may take around 30 seconds per coffee maker. How does the cost of this step compare with the material value, assuming all-inclusive labor costs of 10 to 20 euros/hour?

---

With this last question we should note that theoretically, full disassembly has the benefit that all materials are concentrated with 100% grade and recovery (ultimately, that is what "full" means in this context). In practice, however, certain multi-material components will always contain joints that cannot be disassembled in a reasonable amount of time. For this coffee maker, one such component is the water heating element, which consists of aluminum, copper, some ceramics (for insulation), and some steel, plus a thermocouple switch in its own plastic housing. It may be possible to come up with designs that are better suited for disassembly, but in the current situation, such components will significantly reduce the actual recoverable material value.

---

**EXERCISE 14.16**

Now assume full shredding. What is the cost per coffee maker?

---

**EXERCISE 14.17**

How many such coffee makers would be needed to make setting up a dedicated coffee maker recycling facility economically attractive? How does this compare to your answer to Exercise 14.13?

---

**EXERCISE 14.18**

Adding up the results from the previous six exercises, does it make economic sense to have one or more national recycling plants fully dedicated to discarded coffee makers? Would such a plant involve mainly manual disassembly or mainly shredding?

---

If you have done the previous set of exercises justice, you will understand why smaller products, if they are recycled at all, are usually processed as a mixed stream and using as much automation as possible!

To put all of this in the proper context, first consider that in industrialized countries, the per capita gross national product typically lies around 20,000 euros/year, or about 10,000 euros/year insofar as this is directly related to products, not to services. Of this amount, the actual materials make up roughly 10%—in other words, the average sales value of products is equal to 10 times their material value (with a wide spread of course, much less for "products" such as buildings, much more for medical products such as contact lenses). This immediately shows just how much more attractive re-use is, at the level of products and their components, as compared to recycling at the material level (again, see Figure 14.1). Car disassembly centers do not disassemble parts for their material value, but for refurbishment and re-use (though also to a lesser extent to preserve the material value of the remainder).

More interestingly, even in countries where recycling is relatively highly developed, such as the Netherlands, Germany, and Austria, the value of all recycled materials adds up to just 15% of total material consumption. Expressed this way, recycling still has far to go. Governments have a wide range of measures at their disposal to stimulate recycling. For instance, they can make landfilling and incineration more expensive (via taxation) or even ban it altogether. Also, they can put more responsibility for product recycling in the hands of the manufacturer instead of those of the final owner, and they can impose legal recovery targets for products of particular types. A third option is to put specific taxes on products—"end-of-life fees"—to fund the setting up of dedicated recycling facilities. Faced with the need to secure the supply of materials, governments will increasingly resort to such measures; indeed, in Europe, many such policies and laws are already in place. This puts recycling into a constrained and dynamic marketplace.

---

**EXERCISE 14.19**

Revisit Exercises 14.1 and 14.3. What are your answers now?

---

## 14.5 ECOLOGICAL ASPECTS OF RECYCLING

Recycling processes require energy to operate, and they do not run emission-free. However, compared to making materials from primary sources, recycling consumes significantly less energy and produces fewer harmful emissions. The fundamental reason behind this difference is that recycling only involves mechanical and physical manipulation of matter (shredding, moving, melting), which typically require one to two orders of magnitude less energy than the chemical manipulation involved in primary production (synthesis of polymers, and smelting of metal ores, which may start at a low grade themselves). Table 14.2 gives some indicative values for the "gross energy requirement" (GER) of various materials—that is, the energy needed to make one kg of material for the primary and secondary routes.

---

**EXERCISE 14.20**

Expressed as a percentage, how much reduction in energy does recycling offer as compared to "virgin" production for the five materials in Table 14.2?

However, it is not straightforward to determine if recycling is beneficial in any specific situation. For the secondary GER values found in the literature, it is not always clear which recycling steps have been included. Table 14.2 presents an inclusive view (from collection to reprocessing, see Figure 14.1), but this is not universal, and due care must be taken to prevent unwarranted conclusions. Aluminum presents a key example: the well-known claim that recycling of aluminum "requires only 5% of the energy of primary production" applies only to the re-melting step (as the previous exercise showed you, the actual value is nearly twice as high). Moreover, the comparison ignores the reality that for many materials, primary and secondary production phases are not separate, but closely linked. We already encountered steel as a key example: in "integrated steel production", scrap steel (either post-consumer or production scrap) is used as an indispensable input, keeping the process cool while the surplus carbon from pig iron—the primary input—reacts to form $CO_2$. One result of these links is that in the material market, a significant percentage of the "primary" supply already contains recycled material. And of course, the different metals themselves are linked: aluminum, for instance, is used as a deoxidizer in steel production, whereas silicon is an alloying element in cast aluminum, and so on. The largest usage of magnesium, by weight, is actually as an alloying addition to wrought aluminum alloys!

To add to the complexity, note that GER values capture only a single aspect of the total environmental impact of the material production and can easily lead to a distorted view. For instance, if primary aluminum is made using hydropower and secondary aluminum using coal power, the GER values may tell almost the opposite story about the eco-impact.

To give a different example, the primary production of metals from sulfuric ores (e.g., copper) often causes emission of sulfur oxides, which happens much less during recycling of these metals. Depending on the environmental regulations and emission control that govern processing, this can give huge differences in eco-impact between the primary and secondary routes. Again, the GER values tell only a part of the story. But provided we apply some rigor to the way we combine different types of energy consumption, allowing, for instance, for differences in fossil-fuel share of the electricity supply in each region, quantities such as GER are a good single measure. This follows the recognition by international bodies that carbon emissions are the dominant factor in climate change, with these emissions scaling closely with fossil fuel usage.

---

**EXERCISE 14.21**

Which other environmental impacts can we consider, apart from energy and carbon emissions, when we compare primary and secondary material production?

---

Apart from ecology there is the issue of worker safety. In industrialized countries, recycling is generally quite safe in terms of absolute numbers, but compared to the industrial average, its casualty rate is still several times higher. This is mainly due to accidents in human-vehicle interaction, which are unfortunately inherent to raw materials handling operations in labor-intensive settings. In developing countries, the situation is even worse: there are indications that recycling of, for instance, discarded electronics in certain countries is sometimes done at considerable risk to human health. These issues are, of course, not the responsibility of individual designers but fall under the broader remit of "sustainable development", in which everyone is a stakeholder. The same could be said for many ethical aspects of production, not just recycling—the use of "sweat shops" in developing countries, for instance.

## 14.6 BEST PRACTICE CASE STUDIES

As stated before, most products are not recycled with dedicated equipment but are generally processed as a mixed stream. Exceptions involve simple products made in huge numbers, such as glass bottles or aluminum beverage cans. By virtue of their large numbers *and* the fact that they contain a lot of metals as well as components for re-sale, cars are another exception. In some countries, the same goes for "white goods" (refrigerators and washing machines). Next we describe how these three types of products are recycled; we also show what happens to electronic products and to municipal waste. Note that we focus on best practices as they are being applied today. In the years ahead, it can be expected that such operations become steadily more common and cost-effective.

### Car recycling

With cars containing hundreds of materials and thousands of components, you would perhaps not expect that they can be recycled as efficiently as they are, with up to 85% of all the materials being recycled, depending on the region and location. Then again, it has taken a considerable effort, both technologically and in terms of legislation, regulations, and organizations, to get this far. End-of-life cars are first partially disassembled to secure up to twenty different types of components (e.g., starter engines, batteries, air bags) and materials (e.g., tires, oil, glass: in setting up the necessary infrastructure to collect these materials at high grade, end-of-life fees have been indispensable). Next, the stripped hulk is shredded in a hammer mill and the chunks are fed through an advanced separation setup similar to that shown in Figure 14.2 but with many additional separation and sorting systems. The main metals are recovered very well this way—aluminum up to 80%, copper up to 89%, iron and steel even up to 98%. However, plastic recovery (apart from disassembled parts, such as plastic bumpers) is poor at only 2%. Also, 15% of the input still ends up as "automotive shredder residue" (ASR, or more simply expressed: "fluff"). Despite vigorous research by many car manufacturers, no satisfactory use of this tailing has yet been found.

Note that not all car shredders perform equally well, and what is described here is best-case performance. In the near future, however, we can expect this performance to become more common, particularly in Europe, where it will be mandatory from 2015 that (by weight) all cars are to be recycled with 85% total material recovery (with an additional 10% processed with energy recovery) and, crucially, with the car manufacturer carrying the responsibility for it and thus having a clear incentive to invest in design for recycling.

---

**EXERCISE 14.22**

Composite materials (e.g., used in the BMW i3 electric city car) are not compatible at all with today's car recycling infrastructure. Nonetheless, composite car structures save weight and, hence, energy. Does an 85% recycling target for a composite car make sense?

---

**EXERCISE 14.23**

See if you can find analyses of the energy benefits of lightweight vehicle design, compared to the extra energy expended in making the lighter materials (which may or may not be partially recovered during recycling). What are the main factors that determine the trade-off between the two?

## e-Waste

The consumption of electrical and electronic equipment, such as TV sets and refrigerators, but also smaller household appliances (e.g., steam irons, electric toothbrushes), computers, and cell phones, has been steadily rising. Once discarded, these products turn into so-called "e-waste", and a whole new recycling industry has emerged to deal with this waste stream. Under current best practice, only a few types of products are processed separately on an industrial scale: screens (CRT and LCD), energy-efficient lighting elements, refrigerators, and washing machines. For the last two, the process route is described here, with specific attention paid to material recovery. The many other products (stereo sets, microwaves, cell phones) are processed as a mixed stream of e-waste. This process route is also described here:

1. For refrigerators, step 1 is to secure the CFC coolants (major greenhouse gases!). Next, the compressors and power cables are disassembled and processed separately to recover their copper content. For safety and environmental reasons, mercury-containing switches are removed as well. The machines are then shredded, with an air suction system removing the PUR insulation foam. Finally, the chunks are separated into a ferrous fraction, a non-ferrous fraction, and a plastic fraction (mostly ABS), using the setup shown in Figure 14.2. Overall material recovery is around 85%, ranging between 90% and 99% for the main metals and between 60% and 70% for some of the plastics, with a typical throughput of 400,000 machines per plant per year.
2. Washing machines require no disassembly and can be shredded straightaway, after which the chunks are fed over a separation line similar to the one used for refrigerators. Overall material recovery is around 80%, ranging between 65% and 99% for the main metals and between 50% and 70% for some of the plastics, with a typical throughput of 35,000 machines per plant per year.
3. Mixed e-waste is first manually sorted when it comes in over a conveyor belt, with workers removing all screens, as these require separate treatment. Also, power cables are cut off, again for their copper content. For safety and environmental reasons, components such as batteries, printer cartridges, and mercury switches are removed. Next are shredding and magnetic separation to generate a ferrous fraction, which itself is then hand-picked to remove the copper-rich transformers that would devalue the fraction. Again as in Figure 14.2, this is followed by eddy-current separation and sink-float sorting. The remainder is fed through various other sorting machines (e.g., using color, density, near-infrared light or X-rays) to generate additional plastic fractions. Overall material recovery is 75% to 85%, ranging between 95% and 99% for most metals and 70% for some of the plastics, with a typical throughput of 20,000 tons of e-waste per plant per year.

As always, new developments are just around the corner. For instance, it may be feasible to replace hand-picking of the ferrous fractions with robots: with suitable image processing software, the copper-rich transformers could be sorted automatically, and possibly at a net profit (robots are expensive to install, but relatively cheap to run).

**EXERCISE 14.24**

How do the overall recoveries compare for the three process routes described? How can you explain the relatively high recoveries for arguably the most complex product of them all, the car?

## Waste-to-energy plants

In countries where landfill space is scarce or expensive, municipal waste is often incinerated to reduce volume, and increasingly this is done with energy recovery, generating electricity and heat (e.g., for a regional domestic hot water system). Total energy efficiency can be up to 25%. This is much less than for regular fossil fuel power plants, mainly because waste is not a very good fuel, and because the intense waste-gas cleaning that is required uses a lot of energy. The amounts of waste are vast (in industrialized countries, typically 300 kg per person per year), and consequently the plants are large-scale, with the largest easily processing 1 million tons per year. Around 20% of the input is not burned and comes out as bottom ash: typically, just over 10% of this stream consists of metals (mainly steel, but also aluminum, copper, etc.), the bulk being a chemically inert ceramic residue enclosing the valuable metals. Through subsequent processing, it is possible to recover part of the metals from the bottom ash. The residue itself can be used as filler material in road building but has few other applications.

So waste-to-energy plants perform quite well in terms of volume reduction and energy, but from a material recycling perspective, the result is poor. To improve recovery, metals, paper, and stone can be separated from the waste prior to incineration: this is not only technologically possible but even economically feasible, and it is already done at certain locations. Some of the plastic packaging can also be recovered before incineration; however, this does require end-of-life fees to make it economically attractive.

---

**EXERCISE 14.25**

The heat of combustion of PP is 46 MJ/kg. How does this compare to its GER? What if we also factor in the 25% efficiency? How "good" is "thermal recycling" in this case?

---

## 14.7 CONCLUSIONS: TOWARD REALISTIC DESIGN FOR RECYCLING

Having arrived at the end of this chapter, you, as a designer, are now able to judge the following:

- Whether your product will have a large enough market volume to go for dedicated, separate collection and recycling (e.g., PET bottles, aluminium beverage cans);
- Whether your product will have to be collected and processed along with a number of similar products for reasons of logistic efficiency (e.g., cars, refrigerators, washing machines, with products of different makes put together);
- Whether your product will end up in one of the large mixed waste flows, such as mixed e-waste or household waste.

For all three options, you now also know the typical current levels of recovery of the materials you might select for your product. Furthermore, you will understand that manual disassembly of discarded products only makes economic sense if done by means of simple, low-cost operations, and often only if it allows component re-use instead of the (much less valuable) material recycling. Finally, considering that a product's material value is only a small fraction of its functional value, it should be clear that design for recycling should never compromise product functionality. One interesting school of thought, however, has shown that clever design for recycling can sometimes tap into the perceived added value that the consumer associates with purchasing an "eco-friendly" product.

To learn more about recycling, we refer you to the following sources, as well as to the industry itself: if you ever get the chance, do visit a recycling plant and open your eyes to this normally unseen world of noise and mess, but with remarkably high-grade raw materials emerging at the far end.

## 14.8 **Further Reading**

Academic journals we can recommend are *Resource, Conservation and Recycling*, *Materials and Design*, and *Recycling International*. For car recycling, the leading industry event is the International Automotive Recycling Conference (IARC); for electronics, it is the International Electronics Recycling Conference & Expo (IERCE). Both events are held annually. The following books on the wider issues of sustainability, and recycling specifically, are recommended:

Allwood, J.M., Cullen, J., 2012. Sustainable Materials: With Both Eyes Open. UIT, Cambridge, UK. available as a free PDF for personal use via www.withbotheyesopen.com. (*A readable and informative book concentrating on the "big five" materials: steel, aluminum, cement, plastic, and paper. Using Sankey diagrams to quantify material and energy flows, the book looks at both sides of the problem and its solution: reducing energy use in material production and using materials more efficiently. Highly recommended.*)

Ashby, M.F., 2013. Materials and the Environment, second ed. Butterworth-Heinemann, Oxford, UK. (*A comprehensive overview of all aspects that together shape the complex problem of making eco-informed material choices, ranging from conventional topics, such as energy content and life cycle analysis, to emerging issues, such as material shortage and criticality; includes a wider discussion of sustainable development and the role of materials within it.*)

Ehrig, R.J. (ed.), 1989. Plastics Recycling. Hanser, Munich, Germany (*Old but not outdated, this handy book describes plastic recycling in considerable detail, organized by type (e.g., PET, polyolefins, engineering thermoplastics), with attention to the polymer chemistry involved as well as to the actual equipment used.*)

Reuter, M.A., et al., 2013. UNEP Report, Metal Recycling: Opportunities, Limits, Infrastructure. Accessible via www.unep.org > Resource Panel > Publications > Metal Recycling. (*A vast report well worth reading in full, or simply for its up-to-date tables on the compatibility of different metals in waste streams in the "metal wheel."*)

# Manufacturing Process Choice

# 15

## CHAPTER OUTLINE

## 15.1 INTRODUCTION

We have come to the end of this book and to our last topic of discussion: manufacturing process choice. The final question is how to find the right process for the part or product we want to produce. It is not an easy question—in fact, it is one of the most complex issues to come up in product design, because it strongly affects how a product looks and feels, how much it will cost, and whether it can function as intended. Furthermore, the question rarely has a single answer: a number of different processes can usually be used to form the chosen material into the same shapes. Note too that the detailed choice of which material variant to use is also refined by choosing the process, and by the processing parameters. So process choice is a multi-faceted optimization problem, and there is no single measure of what is "best" in this context. We make no claim that there is a simple formula to provide all the answers, but there are systematic points to guide the decision-making process, which we will present here. The discussion is of necessity quite abstract, as it must be applicable to a wide variety of possible products, materials, and processes. Selected examples are provided to illustrate particular situations, the diversity, and the type of thought process required.

In this chapter you will not find any exercises: choosing processes is something you need to experience for yourself during design projects, and the options are too diverse to be meaningfully captured in this format. We will re-encounter, however, the two common threads that bind this book together: (1) distinguishing among the abstraction levels of principle, method, and equipment; and (2) taking into account the trade-offs within the process triangle of function, quality, and cost. As we shall also discover, the question is not just *how* to choose the right process, but also *when* to do so. Other literature on process choice does not always recognize this issue, but here we will give it the attention it deserves. To this end, we first discuss the product design process, setting the stage for how to approach process choice. Two case studies then follow to make everything come to life, and we conclude with suggestions for further study, as usual.

## 15.2 WHEN TO CHOOSE: THE PRODUCT DESIGN PROCESS

New products do not simply appear out of thin air, but result from a well-established process that delivers results on time and within budget—or at least, a process that strives to attain this ideal. This product design process consists of several consecutive phases, usually with a degree of iteration among them. Exactly which phases there are—or indeed, how many—is a source of some debate; here we distinguish four phases that design theorists generally recognize:

1. Fuzzy front end
2. Conceptual design
3. Design engineering
4. Detail design

Phase 1, the "fuzzy front end", is the beginning. It is called "fuzzy" because, more often than not, the design process emerges organically from wider company activity. For instance, a company may gradually discover that its product range is becoming outdated and decide to hire an external design agency to come up with something new. Alternatively, several companies may decide to jointly explore new possibilities (e.g., the household electronics manufacturer Philips and the DE coffee company joined forces and developed the successful Senseo machines with coffee "pads"). Or a company may decide to transform a certain new technology into an actual product. The exact "when" of the beginning can usually be debated, as can the "why" and "how". But however fuzzy the start may be, the end of this phase should always be concrete: a clear specification of requirements for the product to be designed or, in other words, the definition of the problem that the design process must solve. This is the input for the next phase.

Phase 2, the conceptual design phase, is when the first solutions are conceived and explored. If we are dealing with so-called "normal design", then both the working principle and the general layout of the product can be carried over from existing products or prototypes. As the name suggests, normal design is quite common, and most design can therefore be considered as redesign. For an example, think of a household appliance company adding a new model to its existing range of vacuum cleaners. Alternatively there may be a proposed step change in the design, and the product's working principle or its general layout (or both) are unknown, in which case we are dealing with the (much rarer) "radical design". Examples would then include the Roomba robotic vacuum cleaners, James Dyson's cyclone design, or, indeed, the first electric vacuum cleaner itself from around 1900. Whatever the situation, conceptual design is concerned with proving that the product's requirements can be met—that is, by generating a design proposal that will actually work and that can be produced at acceptable cost—in short, a *design concept*. (By this definition, many "conceptual products" put forward as such by industry are actually less than real concepts, because they often do not work and are certainly not affordable!) Conceptual design relies on creativity and synthesis as much as on calculation and analysis: defining usability aspects, deciding on style and image, making the preliminary choice of materials, choosing which components to make and which to buy in, estimating the target price, and so on. In this phase, analysis is broad rather than deep— the aim is to choose the most promising concept from a range of options, not to refine a single one at the expense of all others. The phase generally ends with a presentation of the design concept to senior management, who decide whether or not to assign more development power to the process (more concepts die than you might think) and to proceed with phase 3: design engineering.

In this third phase, the chosen design concept is elaborated and optimized. Analysis now gets narrower and deeper: rules-of-thumb and back-of-the-envelope calculations make way for rigorous computer-aided design, and systematic prototyping and validation by testing become essential. Creativity remains indispensable, but it is now applied to generating small, practical solutions rather than coming up with grand designs and new vistas. The underlying aim for this phase is simple: to eliminate (or failing that, minimize) all risk. After all, this is also the phase in the design process where the main product-specific investments are ordered, such as dies, molds, assembly jigs, and so on, and simple prudence requires that this money is well spent.

Finally, in phase 4, the whole design is fully detailed, documented, and carried over to the departments responsible for manufacture and marketing. At this stage, any changes made to the design should generally be small and tend to be related to these two domains.

Note that in normal design, it is customary to start preparing for manufacturing and marketing much sooner: why wait if you already know what you will be making and selling, and when you will do so? Indeed, computer renderings of products can appear in sales brochures when the actual products do not even exist yet! In radical design, such "concurrent engineering" is generally impossible, simply because radical innovation cannot be rigorously planned.

With this background, we can now state something of great importance and near-general validity: *There is no single moment in the design process when the manufacturing choice should be made.* Instead, this choice is made gradually and iteratively, defining the manufacturing processes more and more as we progress. Manufacturing options first enter during phase 2, conceptual design, when we should at least decide on the principles we need for manufacture (e.g., for a metal component, casting or forming; for a plastic component, injection molding or thermoforming). We may begin to consider the methods (e.g., sand casting or die casting, standard- or 2K-injection molding), but this determination principally takes place during phase 3, design engineering. Then in phase 4, detailed design, it is time to select the manufacturing equipment, including a consideration of processing conditions, which means that we are now talking closely with suppliers. Figure 15.1 depicts these choices against phases 2 through 4 of the product design process as discussed earlier. In normal design, many of these choices can (and often are) carried over from previous product generations, but this need not, and indeed should not, always be the case: too much of that would stifle innovation.

And what about phase 1? Any manufacturing process choices there? Generally, no: because the fuzzy front end is concerned with defining the "problem" rather than its "solution", there is nothing yet to choose. An exception is when a company decides during this phase to find new applications for its existing (perhaps newly invented) manufacturing equipment. However, one could then argue that the choice is simply part of the program of requirements and is no longer a variable.

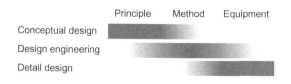

**FIGURE 15.1**

Process choices versus product design process phases. (For color version of this figure, the reader is referred to the online version of this chapter.)

So the choices are to be made step by step as you work your way down from principle to method to equipment. This ensures that at each phase, manufacturing gets proper attention without ever becoming too dominant. In this sense, it is analogous to how you should select materials: from very generic in phase 2 (e.g., aluminum alloy or low carbon steel), to more refined in phase 3 (e.g., aluminum alloys from the 5000 or 6000 series), to even more detailed in phase 4 (e.g., alloy 5054-H2 or 5054-H4). The shape of your product too is gradually defined and optimized, from generic sketches through to detailed engineering drawings, as you specify every dimension, tolerance, and finish. So now let us look at the "how" of choosing the right manufacturing process alongside our choice of material and shape.

## 15.3 HOW TO CHOOSE: SHAPE, MATERIAL, AND PROCESS

Making gradual choices allows you to use the chosen manufacturing processes as additional input for the design process itself. This is of essential importance, as the shape, material, and process of each of your design components all influence each other. As a first illustration, consider the three bicycle frames shown in Figure 15.2. Nominally, they all serve the same basic purpose, but the materials are very different and so are the manufacturing processes used for their constituent components, as are their shapes. Aluminum can be readily extruded into tubes, which can then be welded together and finished to make a typical diamond-shaped frame. Magnesium cannot be extruded nearly as well as aluminum, but it is considerably easier to cast, and provided its shape is suitably changed, a magnesium bike frame can be cast as one piece. Carbon fiber–reinforced plastics (CFRP), in turn, are best here as a thin sandwich structure, produced by resin transfer molding. For good measure, we should add that diamond-shaped frames have also been made from magnesium and composites (largely to meet certain quaint requirements that apply to professional cycling), but such shapes are actually suboptimal for both materials. In fact, it is a common experience to find that the first generation of design in a new material replicates the shape in the old material, and only later are more radical solutions proposed that truly exploit the benefits of using the new material.

For a second example, think of the aluminum space frame developed by Audi for its 1994 A8 flagship model to compete with automotive steel unibodies (Figure 15.3). Choosing aluminum to save weight opened the door to two processes that are not suited to steel: casting and extrusion. Because press forming of sheet aluminum offers somewhat less form freedom than the same method applied to sheet steel, these parallel manufacturing opportunities were essential. Just as for the bicycle frames,

**FIGURE 15.2**

Dependency of shape, material, and process illustrated with three bike frames: aluminum tubular frame, extruded and welded (left); cast magnesium frame (middle); CFRP frame made by RTM (right). (For color version of this figure, the reader is referred to the online version of this chapter.)

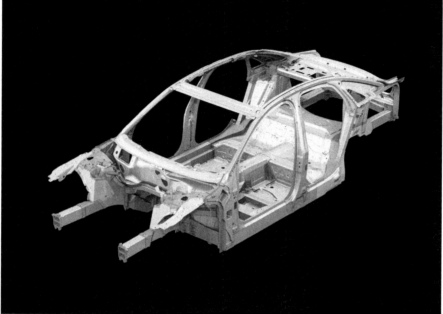

**FIGURE 15.3**

The 2010 Audi A8 and its all-aluminum space frame (red: castings, blue: extrusions, green: panels). (For color version of this figure, the reader is referred to the online version of this chapter.)

the shape of components was significantly affected: for instance, a suspension mount in sheet steel looks quite different from one made in cast aluminum, even though the function is the same.

To give another illustration, consider that injection molding and thermoforming can both be used for plastic housing components. But the former allows a wider choice of materials and has more form freedom than the latter, albeit at the expense of higher investments. It also has closer tolerances, allowing the use of, for example, injection molded housing shells as assembly trays for the many other components in your product. Thermoformed housings usually do not allow this convenient assembly solution, requiring instead some kind of separate frame to keep everything in place.

Discovering how a certain manufacturing process makes sense—that is, how it naturally combines with the necessary shape and material—is not easy and is to some extent a matter of experience. Nevertheless, it forms the very heart of the design process. For the first pre-selection, a systematic screening process has merit, because it eliminates processes that cannot meet one or more of the design requirements, such as the material class or the proposed component size or production volume. This is the basis of the methods developed by Ashby and co-authors, implemented in the Cambridge Engineering Selector software (see Section 15.7 Further Reading for further details). Automating the process in this way has its limits, however. A database of the capabilities of every process can only give broad inclusive ranges for the attributes—it is impossible to include different data for every process-material combination, and there will be many coupled interactions among shape, material, and process that are too complex to capture in simple data ranges. Rather than providing detailed solutions, the principal benefit in using these methods is perhaps to provide a "checklist" for inexperienced designers that introduces them to the issues they should be considering in design. Designers need to accumulate experience and awareness of what works in practice, and to have their eyes open for common pitfalls, as they iterate through the product design process. So a thorough understanding of what each process can do to each material is essential to avoid repeatedly heading down a dead end, and this is why Chapters 3 to 12 of this book are so important. Shape and material and their inter-dependencies with process are central to the problem, but recall that there are other variables that affect manufacturing process choice. We have seen them before: they are the other attributes presented in the manufacturing process triangle. Now it is time to take a closer look.

## 15.4 DIGGING DEEPER: THE PROCESS TRIANGLE REVISITED

Shape and material are two of the main drivers for manufacturing process choice (either on the level of the principles or that of the methods), but there are the three other functional attributes presented in the process triangle, plus all of the quality- and cost-related attributes. Recall that, in principle, all fifteen have attributes inter-dependencies. Here we will focus our remarks on the most influential attributes and their interactions and trade-offs—the process chapters provide additional examples and explanations.

### Function: beyond shape and material—size, production volume, and "look and feel"

The first of the other functional attributes to consider is size. Obviously, the process should have the capacity to produce the part or product in the required dimensions, volume, or weight. Different methods under one principle generally have different practical limitations (e.g., sand castings can be much bigger than die castings). Note also that if you move to the extreme ends of what is technologically possible, you will have very few suppliers to choose from, as the necessary equipment will then be scarce (and expensive, an immediate trade-off with another attribute). Alternatively, if one process cannot handle the required size, it is usually possible to build up a component out of several smaller parts. Largely for this reason, size is important, but not crucially so, for process choice—at least during conceptual design.

Our next attribute, production volume, is much more influential: in fact, together with shape and material, the annual production volume is often decisive for process choice, even at the conceptual design phase. With the exception of one-off products and prototypes, we always want to produce more than one object (e.g., 50,000 vacuum cleaners per year for sale in a certain region, for a projected production run of five years) and the process we choose must be able to do so. Approximate "typical" ranges of production volumes that a process method can handle have been estimated. Assessing the viable production volume is effectively a surrogate for "economic", without going into the actual costs themselves. But in practice this attribute may be highly dependent on what kind of equipment is being used (the degree of automation, for example), so taking a closer look is essential. Note also that especially regarding this attribute, flexibility is good. After all, a process that can handle a wide range will still be okay if your actual sales figures are much different from what you expected, whereas a process with a narrow range would then cause serious financial trouble.

Concluding the functional attributes, we arrive at "look and feel": the color and texture that a part should have, the impression it makes on the user, surface details such as patterns, logos, or text, and so on. Important as all of this may be, this attribute rarely is a decisive factor for process choice, as it is partially implicit in the material choice you made. Moreover, the desired look and feel can also be achieved through surface treatments, such as painting or printing. Still, it is definitely a bonus if such extra steps can be avoided. Molding of plastics stands out as the principle that has the most to offer in this respect; however, it is always worthwhile to see what a process can add to your product, even if no specific look and feel is required (Section 15.5 presents an example where this added benefit turned out to be very important).

## Quality: tolerances, roughness, defects, properties, reproducibility

The first of the quality attributes concerns tolerances. Parts and products obviously need a certain shape to function, but the question is, how precisely can this shape be realized? Are two parallel holes indeed perfectly parallel? Is that right angle exactly 90 degrees? How close is that cross-sectional area to the design (i.e., will the applied stress be as intended)? And so on. Of course, perfection is never possible (or necessary). There always exists a small deviation from the target, and tolerances describe how much of a deviation can be tolerated (hence the term) without the part losing its functionality. Generally, this aspect of the triangle does not directly drive process choice, particularly as suitable secondary operations can often bring tolerances under control (e.g., first cast, then machine; or first stamp, then edge trim). Instead, it is one of the first aspects to consider once a preliminary choice has been made (e.g., based on shape, material, and part price), and then the shape is adjusted accordingly if it appears that certain tolerances will be too wide. Still, it is an important bonus if a process has such intrinsic accuracy that extra steps are not needed, thereby reducing costs. Pressure die casting of magnesium and sintering of steel come to mind, as does the ubiquitous injection molding of plastics as well as integrally machined "unibody" design, such as the Apple iBook. Of course, for materials that are hard to process with such secondary steps, "net shape manufacture" is a must.

Here it is particularly important to spell out certain inter-dependencies. Tolerances are strongly influenced by three of the other attributes in the triangle, namely size, shape, and material. This rule is perhaps best illustrated with the injection molding process. It will be easy to mold a cell phone housing with a length of $80 \pm 0.2$ mm, but the same tolerance will be nearly impossible (and completely unnecessary) for a 1.5 m long garden lounge chair—and will certainly be very expensive. So the larger the part or product, the wider the tolerances should be. Shape also plays a role: simple, easy-to-mold shapes with predictable flow patterns will give better tolerances than complex, hard-to-mold shapes.

Likewise, material comes in: due to its much smaller shrinkage, the amorphous ABS typically used for cell phones will give better tolerances than the semi-crystalline PP applied in outdoor furniture. Residual stress is another source of dimensional variation, and this depends on the cooling history and hot strength—which are, once again, sensitive to the shape and material. These are just examples; significant trade-offs among tolerances, size, shape, and material exist for all processes.

Moving on to our second quality attribute, recall that roughness defines how smooth the surface of our part or product is. It is generally expressed as the average height, in micrometers, between the peaks and troughs of the surface, seen at the micro-level and usually now measured with suitable laser equipment. Because after manufacture, roughness can only be affected by relatively expensive surface treatments (e.g., polishing), it is important that the roughness inherent to a process is acceptable—or that the proportion of the part that needs secondary finishing is small. Certain principles stand out as being able to offer a wide range of roughness values. Again, injection molding comes to mind (consider the difference between smooth and textured molds), but also machining. Note also that roughness can vary strongly among methods under the same principle (recall how sand casting gives a much rougher surface than die casting). Generally speaking, roughness is not a main driver for process choice, but it is definitely a factor of importance for the majority of everyday products. For fatigue-sensitive parts, roughness *can* be decisive: the smoother the surface, the longer that fatigue crack initiation can be delayed.

Defects can be broadly subdivided into those within the bulk and those at the surface. *Bulk defects* cover factors such as porosity, internal microcracks, or major inhomogeneities in microstructure that mostly influence mechanical properties. This is of crucial relevance to safety-critical parts that must function under high loads, such as metal car suspension components. For such parts, ordinary casting methods are unsuitable (because of the inherent casting defects, such as porosity) and, instead, forging is the usual choice, despite the higher cost and somewhat smaller form freedom compared to casting. (Audi, in case you are wondering, uses another alternative: casting under vacuum.) *Surface defects* often relate to local quality issues in processing: parting lines in castings, breakout edges of stamped metal parts, burrs on machined components, "cold shut" and quench cracks in forging and heat treatment, knit lines and ejection marks on moldings, and so on. Defects are not normally a strong driver for process choice, but they need to be understood for each principle so that they can be avoided or managed by the design and the processing conditions. For example, some imperfections can be removed, or covered up with a coat of paint, or, even better, hidden by clever design, whereas others can be eliminated by attention to detail (such as avoiding sharp changes in section thickness). But apart from deploying secondary operations or making design changes, it is often the chosen equipment and supplier that can control defects: for instance, high-tech manufacturers who know exactly what they are doing tend to give a much better result than those who are mainly concerned with speed and bulk. This obviously goes for the other quality attributes also, but nowhere can you literally see the result of good choice of supplier and equipment.

Turning next to material properties, processing plays an important role in developing some, but not all, of the properties of the product, via the internal microstructure. The stiffness and density, for instance, are mostly insensitive to how the part is made—the packing of atoms is physically prescribed in metals and ceramics. The molecular structure of polymers gives scope for greater variation in stiffness, particularly in semi-crystalline or cross-linked polymers, when changes in cooling history can lead to different outcomes. Fiber composites have moduli determined by the polymer matrix and reinforcing fiber fraction, with the opportunity to customize the Young's modulus by aligning the fibers along load-bearing directions. Where processing plays a really significant role is in determining strength, ductility, and toughness, particularly in metals. Processing governs solidification, material flow, deformation conditions, heating and cooling rates, and so on. These in turn control many microstructural factors in metals that affect these mechanical properties (as well as other structure-sensitive properties, such as thermal conductivity, electrical resistivity, and susceptibility to corrosion). These include grain size, whether the alloy is work hardened or recrystallized, the phases present, and their size and spatial distribution—all of which may be described by detailed theories of phase transformations, dislocation behavior, and diffusion. From the perspective of the designer, we only require an awareness that properties can be quite different in practice from the values found in a database, so it is essential to consult experts.

Closely related to properties is reproducibility (although controlling variability applies equally to dimensions, surface finish, and so on). The question is not just to produce the right quality once, but to do so repeatedly over the full production volume. The gist of the problem is captured in Figure 15.4, which shows the probability with which a given outcome (such as strength) can be achieved by two competing process routes. The wise designer goes with option A—in spite of the lower average outcome, it is the low-quality tail that matters and determines the outcome that can be delivered reliably. With the rise of "total quality management" and the so-called "six-sigma" approach, the issue of reproducibility has been turned into a veritable science, reaching beyond the manufacturing process to influence the design process. We gladly leave this Pandora's box shut and refer to the literature instead, but our discussion of process aspects would not be complete without its mention. Like several other quality attributes, reproducibility is not a significant factor on the level of principle or method, but it is mainly determined at the level of the equipment and, especially, the supplier. So in your process choice, look for features such as ISO 9000 quality certification, six-sigma credentials, and so on.

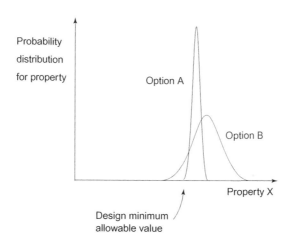

**FIGURE 15.4**

Probability distributions for a key property, as delivered by two processing options. (For color version of this figure, the reader is referred to the online version of this chapter.)

## Costs: part price, investment, speed, time-to-market, social factors, and ecological costs

The first of the cost attributes is the part price—that is, the all-in financial cost of producing one unit for your part or product. As Chapter 8 on injection molding showed with a worked example, cost per part is determined by the material cost; the part-specific fixed investments (amortized over the production volume); the amount and cost of labor, energy, and machine time; the rate of production; and packaging and transport. Some of these factors can also impose constraints in their own right—for example, the production rate must meet the required rate of supply needed (although there are many ways to speed up manufacture, such as by using multi-cavity dies or by having parallel facilities alongside a production line, for jigging, clamping, and other components involved in setting up—although this then increases the level of investment). But in the end, it is the part price that determines if the process is suitable, and for this reason alone it requires special mention. In fact, part price demands strong attention during every phase of the design process, no exceptions.

Regarding cost price, it is again important to spell out a few major inter-dependencies. Selecting relatively expensive materials (a functional attribute) is a common mistake that drives up cost. Often it is better to use, for example, chrome-plated carbon steel instead of stainless steel, or to cover a commodity plastic with a nice foil (e.g., using in-mold decoration) instead of using a high-price engineering plastic for the full part. Likewise, overspecification of tolerances (a quality attribute) will needlessly increase cost. Choosing smart joining and assembly solutions can negate the need for high-precision parts, again saving cost—which is a major reason to address assembly just as early on in the design process as you would address part manufacture.

Next, consider investment, another influential driver for process choice. Although the capital outlay contributes to the part price, it merits special consideration in its own right, as investments need to be paid up front. Small companies launching new products often cannot afford high-investment processes (injection molding, matched die forming, etc.) and must look for alternatives, especially if it is uncertain how many products they will actually sell. For bigger companies producing more established products with reasonably solid sales predictions, this attribute presents less of a problem, as the investments can then be assumed to be amortized over the full production run. Still, no company enjoys sinking shareholder money into a die or mold, so the issue always plays a role; in fact, the investment level can effectively determine the series size and with it, the pace of real innovation—not the other way around (the automotive steel unibody is perhaps the best example of this effect). And once again, the investment levels vary strongly among different principles, different methods under one principle, and different kinds of equipment under one method. Note that investment not only includes "hardware" but also the actual design and development process: as a designer, you wish to be paid just as much as anyone else!

Moving on, we get to time-to-market: the time needed to design the part or product, to arrange the necessary investments, and to set up manufacture, leading all the way to "job one" (i.e., the production of the first unit, which can then be sold to a customer). Limitations on this attribute are generally more of a barrier to adopting something radically new than are the designers' intentions. Finding new suppliers, for instance, often takes more time than is available, and process choice at the equipment level (and from there up to the levels of method and principle) is then limited to what the designers know, or rather, with whom they have an established relationship. In design practice, this attribute plays a much stronger role than many people think! You should also be aware that especially at the method level, processes differ strongly in the typical "lead time" for making dies, molds, jigs, and so on: from a few days to make part-specific fixtures for integral machining, to several months for sets of dies for matched die forming (additive manufacturing, of course, has a unique advantage here, and this is where the expression "rapid manufacturing" comes from). A rule-of-thumb is that the higher the investment, the longer the lead time. All of this also goes for making changes to parts and products, which some processes allow quickly, whereas others do not.

Concluding the cost attributes, there are the social factors and ecological costs. Chapter 1 set these within the wider context of sustainable development, and the "three capitals": manufactured, human, and natural. These attributes are difficult to incorporate at the level of choosing a process for a design—they are bound up with everything that the company does and its overall policies with regard to markets, employment practices, and so on. In the developed world, workers are protected by a whole raft of legislation governing health and safety, working hours, pay, and conditions, and industrial accidents have become rare. The same cannot be said, however, for industrial environments in developing economies, where safety standards are poor, and women and children are exploited for subsistence wages—so there are ethical issues to be considered in sourcing materials and parts from overseas. At the level of the specific equipment (i.e., suppliers), there can be a lot to choose. Not all suppliers are equally effective in ensuring good working conditions, nor are they equally dedicated to reducing pressure on the environment. So when the time comes in the design process to choose suppliers, ask to see, for example, ISO 14000 certification (which addresses the supplier's environmental management procedures), inquire about worker safety, and incorporate the answers in your decision-making.

### A quick recap

From the fifteen attributes captured in the manufacturing process triangle, it is not just shape and material that need to be considered along with manufacturing process choice, but especially also the production volume and the cost price, regardless of what phase of the product design process you are in. Most of the other aspects are generally less influential, as they can often be solved at moderate cost (e.g., subsequent machining to solve a tolerance issue) or only come in effect later during the design process (e.g., supplier choice). Still, each attribute matters to someone at some point, so consider them all. Tools exist to facilitate this practice, with the *Cambridge Engineering Selector* being a prime example, but eventually you will need to rely on a solid base of understanding.

## 15.5 TWO CASE STUDIES ON MANUFACTURING PROCESS CHOICE

With all of this theory behind us, we now present a few case studies to illustrate how manufacturing process choice can take place in practice. The two we have chosen here both come from the authors' personal experience, so we can guarantee that this is in fact how things took place. One note of caution: they are examples of particular situations and do not necessarily incorporate general rules of process selection (practice rarely follows theory that perfectly).

## Lightweight extrusions in heavyweight trucks

Based in Eindhoven, the Netherlands, DAF Trucks is one of the leading players in the European commercial vehicle industry, annually producing some 35,000 trucks in the medium- and heavy-size markets (equivalent to Class 8-9 trucks in the United States). The "space cab" version of DAF's CF85 model (Figure 15.5) was originally designed for national transport and did not therefore include two full-size sleeping bunks. Around the year 2000, DAF discovered that this truck was increasingly being used for international transport, so it was decided to offer a second bunk as an option for this model, placed high in the cabin. The designers tried to adapt the existing upper bunk from the bigger XF95 truck, but they soon discovered it would be too heavy. Bunk weight was clearly not an issue for the truck's fuel performance, but it was decisive for convenience of assembly: maneuvering a 30 + kg upper bunk in place in the comparatively narrow CF85 cabin would require specific, costly solutions that could not be justified for an optional product.

In a effort to save weight, the heavy steel bunk frame was redesigned in aluminum. However, the first design was disappointing: straightforward replacement of steel with aluminum turned out to save little weight and cost far too much. However, a closer look at the available manufacturing processes offered an excellent alternative. By choosing extrusion, it was possible to put the material exactly where it was needed, saving weight and cost. More importantly, the potential to integrate assembly functions into the profile, such as those for the bunk floorboard and its upholstery, kept total cost under control (Figure 15.6). The alloy in question was 6060-T6: sufficiently strong for this application, yet very easy to extrude, maximizing the design freedom.

For the rear part of the bunk frame, it turned out to be possible to replace the heavy (and squeaky) steel hinges that connect the bunk to the cabin rear wall with rubber hinges that could be fully integrated into the profile, again saving weight. And although this was not part of the initial design brief, the

**FIGURE 15.5**

DAF CF85 medium-sized truck, space cab version. (For color version of this figure, the reader is referred to the online version of this chapter.)

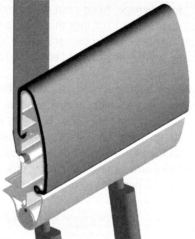

**FIGURE 15.6**

New CF85 upper bunk (top) and its extruded aluminum front profile (bottom). (For color version of this figure, the reader is referred to the online version of this chapter.)

bunk's look and feel was significantly improved as well by exploiting the high-tech appearance of anodized aluminum—an attribute that since then has become common for many other parts on DAF's trucks. Furthermore, aluminum extrusion is well-suited to the production volumes typical in the truck industry, and the necessary investments are modest (just some 4,000 euros for the front profile). All in all, this product is a prime illustration of how function, shape, material, and process can go hand in hand, each reinforcing the other factors.

We should add that DAF did not develop this new bunk alone, but received valuable assistance from its interior supplier Berco in Schijndel, the Netherlands, and from its aluminum profiles supplier Hydro Aluminum in Raeren, Belgium, from the conceptual design phase onwards. So for this product, equipment and suppliers were chosen already very early on. (Such cooperation is common in many sectors, not just the automotive industry.) And to be really complete, we must own up to the fact that initially, the CF85's upper bunk was not the market success everyone had hoped for—but it did pave the way for a new and successful aluminum bunk in DAF's bigger XF105 model, plus many other extruded products. Even with the ideal process choice, success can take some time!

## Coin storage boxes: hidden design jewels for the happy few

Many people collect coins for a hobby—and some people do so for a living. Like comparable museums in other countries, the Royal Dutch Museum's coin department (in Dutch: *Koninklijk Penningen Cabinet*) collects sample specimens of all kinds of coins from modern to ancient. To do this in a neat, well-organized manner, some kind of storage facility was needed. In 1995, Studio Ninaber was called in to design a solution, and soon found the museum's requirements to be surprisingly stringent. For instance, because the coins have to be preserved for the future literally in mint condition, the use of any plastic or foam that releases solvents, no matter how tiny the amount, was ruled out to eliminate any risk of tarnishing the coins. Furthermore, the new design had to allow storage, and handling, of the coins individually, to prevent any damage from coin-to-coin contact (the collection includes, e.g., ancient Greek golden coins that are just 0.2 mm thick and *very* fragile). Space was also constrained, so small coins could not simply be put in the same space as larger coins. And because there are only a few large coins and many smaller ones, these last two requirements combined to point to an incompatibility of series sizes.

Step 1 of Studio Ninaber's solution was to put each coin in a shallow box of its own with a soft, non-slip inlay, space to insert a label, and a clever shape for easy handling: the boxes could be flipped up by pressing down on the label, without touching—and hence, tarnishing—the coins themselves (Figure 15.7). No fewer than 10 different box sizes were designed, ranging from $32 \times 70$ to $194 \times 211$ mm, to accurately match the wide range in coin diameters and minimize storage space.

Step 2 was to select not one, but two different manufacturing processes. The smaller boxes would be made by injection molding: of these, the numbers were sufficient to amortize the high investment for

**FIGURE 15.7**

Coin storage solution for the Royal Dutch Museum.

the molds. The larger boxes, of which much fewer were needed, would be made by thermoforming, which is better suited for smaller production volumes.

Step 3 was to develop special molds based on sintered aluminum. Being porous, these innovative molds had no need for the vent holes that are normally placed in thermoforming molds. This way, the product side that makes contact with the mold could become its visible side, giving sharply defined radii and texture but without unsightly vent marks: in short, getting a high quality look and feel out of a low-investment process. Next, the smaller injection-molded boxes received the same detailing and finish as the larger thermoformed ones. In effect, all boxes ended up looking the same except for their different sizes, giving a coherent and pleasing aesthetic. Finally, a grade of ABS was selected for all boxes with graphite as a pigment instead of the usual metal oxides. This, combined with acid-free paper for the labels and plasticizer-free PP inlays, resulted in a 100% tarnish-free storage solution, meeting the highest international standards for the conservation of art. Studio Ninaber's design was a success and is still in use today, with some 100,000 smaller and 10,000 larger boxes produced so far!

You may wonder, what *is* this product, really, and why present it here? After all, with the exception of the people working at the museum, no one ever encounters these boxes, no matter how well they are designed. But there is a good reason for showing them. Many professional designers work on such niche products that go unnoticed by the masses but that can make a real difference to the people who encounter them, handle them, and work with them every day. The same often goes for most mechanical engineers and the applications they work on. And unlike mass market products, such as cars and cell phones, which are designed by large teams in which each designer is little more than one cog in the machine, these niche products are designed entirely by small teams. So if you really love design and every part of it, you will learn to appreciate niche products such as the coin storage boxes, and the clever use of manufacturing processes behind them.

## 15.6 CONCLUSIONS

This final chapter has turned your attention to the issue of choosing the right manufacturing processes during product design. We have shown that this is not just a question of *how* to choose, but also a question of *when*. Section 15.2 revealed that there is no single point in time when "the absolute" choice is made, but rather, that there is a series of choices, moving steadily downward (and often iterating upward again, if an initial choice leads to a dead end) from principle to method to equipment and supplier during the design process.

You have also seen how manufacturing process choice has to be done jointly with choosing shape and material. Design is not choosing a shape first, then "sticking a material onto it" and eventually seeing if the combination can somehow be made (or worse, expecting that someone else will find out for you if it is). That only leads to disappointment, lost time, and much-too-expensive products, if they can be manufactured at all. Instead, these three elements have to be considered together. It bears repeating: *discovering which combination of shape, material, and process "makes sense" in the light of the required functionality lies at the heart of design*. Beyond shape and material, two additional attributes of the manufacturing triangle come into play at every phase of the design process: production volume and part price—the former in particular has a major influence on manufacturing process choice. The other attributes of the triangle all play a role and are often coupled to the shape, material, and process, whereas in specific situations any of them may become a main driver for process choice.

## 15.7 Further Reading

This chapter does not claim to give the final word on the issue of manufacturing process choice, but only to set the scene within a systematic framework for students of industrial design, mechanical engineering, or a similar program. For further reading, we recommend the following sources:

Ashby, M.F., 2011. Materials Selection in Mechanical Design, fourth ed. Butterworth-Heinemann, Oxford, UK. (*The full exposition of Ashby's selection methodology, for more advanced bachelor's- and master's-level students.*)

Ashby, M.F., Johnson, K., 2010. Materials and Design, second ed. Butterworth-Heinemann, Oxford, UK. (*An accessible book for bachelor's students and professionals alike.*)

Ashby, M.F., Shercliff, H.R., Cebon, D., 2013. Materials: Engineering, Science, Processing and Design, third ed. Butterworth-Heinemann, Oxford, UK. (*A design-led undergraduate text on materials that introduces Ashby's well-known methodology for selecting materials and manufacturing processes and illustrates the effects of processing on properties.*)

Cambridge Engineering Selector, CES EduPack software (annual releases). Granta Design Ltd., Rustat House, Clifton Road, Cambridge, UK. (*This package embodies the Ashby selection methodology for materials and processes from the books listed here, including databases at several levels for all materials and specific industrial sectors. It also includes a life cycle analysis tool for assessing the environmental impact of materials.*)

Industrial Designers Society of America, 2003 Design Secrets: Products—50 Real-Life Projects Uncovered. Rockport, Minneapolis, United States. (*This book presents the secret world of product design for many more products, ranging from the common to the unknown. Highly recommended!*)

Otto, K., Wood, K., 2000. Product Design. Prentice Hall, New Jersey, United States. (*A good all-round book, capturing many if not all aspects of the subject.*)

Pahl, G., Beitz, W., Feldhusen, J., Grote, K.H., 2007. Engineering Design: A Systematic Approach, (translated and edited by K. Wallace and L. Blessing), third ed. Springer-Verlag, London, UK. (*A thorough and detailed treatment of design, aimed primarily at the embodiment design phase.*)

# Index

Note: Page numbers followed by *b* indicate boxes, *f* indicate figures and *t* indicate tables.